SPACE SYSTEMS

Earth, Sun, and Moon
Cyclic Patterns of Lunar Phases, Eclipses, and the Seasons

Derek L. Miller

Cavendish Square
New York

Published in 2017 by Cavendish Square Publishing, LLC
243 5th Avenue, Suite 136, New York, NY 10016

Copyright © 2017 by Cavendish Square Publishing, LLC

First Edition

No part of this publication may be reproduced, stored in a retrieval system, or transmitted in any form or by any means—electronic, mechanical, photocopying, recording, or otherwise—without the prior permission of the copyright owner. Request for permission should be addressed to Permissions, Cavendish Square Publishing, 243 5th Avenue, Suite 136, New York, NY 10016. Tel (877) 980-4450; fax (877) 980-4454.

Website: cavendishsq.com

This publication represents the opinions and views of the author based on his or her personal experience, knowledge, and research. The information in this book serves as a general guide only. The author and publisher have used their best efforts in preparing this book and disclaim liability rising directly or indirectly from the use and application of this book.

CPSIA Compliance Information: Batch #CW17CSQ

All websites were available and accurate when this book was sent to press.

Library of Congress Cataloging-in-Publication Data

Names: Miller, Derek L., author.
Title: Earth, Sun, and Moon : cyclic patterns of lunar phases, eclipses, and the seasons / Derek Miller.
Description: New York : Cavendish Square Publishing, [2017] | Series: Space systems | Includes bibliographical references and index.
Identifiers: LCCN 2016030323 (print) | LCCN 2016036858 (ebook) | ISBN 9781502622914 (library bound) | ISBN 9781502622921 (E-book)
Subjects: LCSH: Astronomy--Mathematics--Juvenile literature. | Eclipses--Juvenile literature. | Seasons--Juvenile literature. | Moon--Phases--Juvenile literature. | Earth (Planet)--Orbit--Juvenile literature.
Classification: LCC QB51.3.M38 M55 2017 (print) | LCC QB51.3.M38 (ebook) | DDC 523--dc23
LC record available at https://lccn.loc.gov/2016030323

Editorial Director: David McNamara
Editor: Caitlyn Miller
Copy Editor: Rebecca Rohan
Associate Art Director: Amy Greenan
Designer: Alan Sliwinski
Production Coordinator: Karol Szymczuk
Photo Research: J8 Media

The photographs in this book are used by permission and through the courtesy of: Cover Markus Gann/Shutterstock.com; p. 4 Luc Viatour/www.lucnix.be/File:Solar eclipse 1999 4.jpg/Wikimedia Commons; p. 10 Don†Hitchcock/donsmaps.com; p. 15 Loon, J. van (Johannes)/File:Cellarius ptolemaic system c2.jpg/Wikimedia Commons; p. 17 NASA/File:Earth precession.svg/Wikimedia Commons; p. 19 Radoslaw Ziomber/File:Total lunar eclipse 2015.09.28 4-15 Rabka-Zdroj.jpg/Wikimedia Commons; p. 21 Dito/ullstein bild/Getty Images; p. 26 Castleski.Shutterstock.com; p. 28 Georgios Kollidas/Shutterstock.com; p. 32 Tau'olunga/Public Domain/File:North season.jpg/Wikimedia Commons; p. 36 Archive Photos/Getty Images; p. 43 Gpalmeric/File:Annular Eclipse, Nevada City, CA - May 2012.jpg/Wikimedia Commons; p. 46 Rvalette/File:Pythéas.jpg/Wikimedia Commons; p. 48 Photo Researchers, Inc/Alamy Stock Photo; p. 54 Stefano Bianchetti/Corbis/Getty Images; p. 62 DEA/G. DAGLI ORTI/Getty Images; p. 66 File:Albert Einstein 1921 by F Schmutzer.jpg/Wikimedia Commons; p. 70 Shooarts/Shutterstock.com; p. 73 Falkenstein/Bildagenturonline Historical Collection/Alamy Stock Photo; p. 76 Tewy/File:Tide table 01.jpg/Wikimedia Commons; p. 80 ESA/Hubble & NASA/File:UZC J224030.2+032131.jpg/Wikimedia Commons; p. 82 Tomruen/File:Lunar eclipse from moon simulation-sep 28 2015.png/Wikimedia Commons; p. 84 John R. Foster/Science Source; p. 89 Jeffrey T. Kreulen/Shutterstock.com; p. 91 Corbac40/Shutterstock.com; p. 96 NASA/JPL-Caltech/T. Pyle (SSC)/File:Massive Smash-Up at Vega.jpg/Wikimedia Commons; p. 100 NASA http://visibleearth.nasa.gov/Getty Images.

Printed in the United States of America

Contents

   Introduction: Cycles on Earth and in Space .. 5

1 Early Predictions 11

2 The Modern Understanding of
   Lunar Phases, Eclipses, and Seasons 27

3 Scientists and Mathematicians 47

4 Visualizing the Movements of the
   Earth, Sun, and Moon 71

5 Lunar Phases, Eclipses, and Seasons:
   Today and Tomorrow 85

   Glossary 102

   Further Information 105

   Bibliography.......................... 107

   Index 110

   About the Author...................... 112

During a total solar eclipse, the ring of plasma surrounding the sun is visible to the naked eye.

Introduction: Cycles on Earth and in Space

Cyclical patterns are common in the natural world. Day follows night, and the seasons repeat themselves again and again. Humans have always searched for patterns in the world, and some of the earliest known writings concern the cycles of the seasons, lunar phases, and eclipses. An understanding of the seasons was essential for humans to master agriculture and grow crops on a massive scale, and agriculture in turn allowed humans to come together in large cities and create civilizations. The lunar phases used to be an important tool for measuring time, and the past importance of these phases can still be seen in the organization of the modern-day calendar into months, which were first based on the phases of the moon. Eclipses were often seen as important omens—and the image of the day suddenly turning to night while the sun is still in the sky can be found again and again in early texts and myths. Some civilizations even reportedly practiced human sacrifice to ward off eclipses. Humanity's early understandings of these space systems will be the topic of the first chapter of this book.

Despite the strides that early civilizations made in astronomy, it is only recently in human history that we have been able to explain exactly why celestial phenomena occur according to fixed cycles. While earlier civilizations were sometimes able to predict them, they could not provide scientific explanations for their occurrence. But we now know that the movement of the sun, moon, and Earth cause natural phenomena, such as eclipses, tides, the seasons, and lunar phases to occur in predictable cycles. And we also know that gravity is responsible for the movement of the sun, moon, and Earth. In this way, it is gravity that drives these observable phenomena.

In chapter 2, we will examine the precise causes of these cycles in depth. We will trace the development of humanity's understanding of these complex astronomical phenomena before turning to our modern understanding and different ways of visualizing these cycles. We will see that the seasons are caused by the tilt of Earth's axis as well as its orbit around the sun, the lunar phases are caused by the sun illuminating different areas of the moon's surface, **solar eclipses** are caused by the moon blocking the sun from sight and causing Earth to fall into darkness during the day, and **lunar eclipses** are caused by Earth blocking the sun's light from falling on the moon—causing the moon to turn red or disappear in the night sky.

We will also look at other ways that the cyclic movements of the sun, moon, and Earth affect our everyday lives. The ebb and flow of the tides were an early mystery that was eventually linked to the orbit of the moon. The creation of a calendar—a seemingly simply invention—was also fraught with missteps and errors. Even today, with the inclusion of leap years and leap seconds, the calendar is still slightly out of step with the movements of Earth around the sun.

In chapter 3, we will learn about the long line of thinkers who shaped our understanding of the solar system. Although they lived in very different times and places, they share some characteristics. They all used evidence to develop theories that explained natural phenomena. Often, their discoveries were at first discounted by the scientific community until experiments proved their theories beyond a doubt. And many of them were also subject to political forces that dramatically changed (and sometimes even ended) their lives. A defining characteristic they share is that they all radically altered the way people understood the heavens. Many of them rose from humble beginnings to great fame due to the scientific advances that they made, and their names—Galileo Galilei, Isaac Newton, Albert Einstein—are now famous around the world.

In chapter 4, we will investigate different ways of visualizing the solar system. Some physical models can be made at school, while others stretch across entire countries. High-tech visualizations, like computer animations and time-lapse videos, are also valuable tools for understanding the lunar phases, eclipses, and seasons. NASA has a rich variety of these tools available to the public on the internet, and we will examine what exactly they reveal about the lunar phases, seasons, and eclipses.

The development of our understanding of the seasons, lunar phases, and eclipses was not a straightforward journey. It was necessary for us to discover that Earth orbits the sun, and not the opposite, in order to explain these phenomena. The early Babylonians and Greeks made great advances in the field of astronomy, but it was not until the scientific revolution of the 1500s that the sun's place at the center of the solar system was cemented. And it was not until Isaac Newton formulated the **law of universal gravitation** in the early seventeenth century that a precise model of the movements of the planets existed, and even then Newton could not tell *why* gravity existed. Only at the beginning of the twentieth century did Einstein offer an explanation of what causes gravity with the development of his theory of relativity.

But even now questions still remain. How will the solar system end? Will the planets collide at some point in the future or are their orbits stable? Will Earth be swallowed by the sun? Or will it continue to slow down until a single day lasts more than one thousand hours? We will examine these questions about the ultimate fate of the solar system in the last chapter of the book.

A firm understanding of the movements of Earth and the moon will help us answer these questions about the future of the solar system. We can see how the lunar phases, eclipses, and seasons will gradually change as Earth's and the moon's orbits shift over time. The one constant will be the cyclical nature of these phenomena, until one day they cease to exist at all. Luckily, that will be millions of years from now and not within our lifetime.

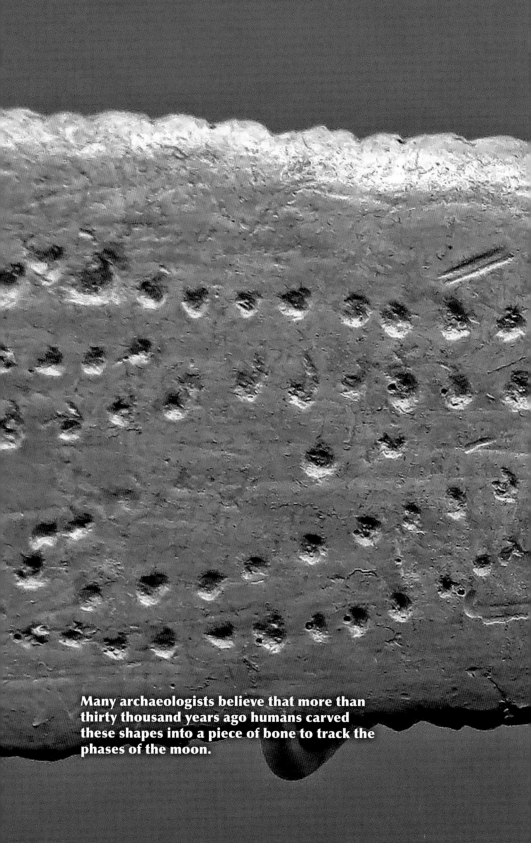

Many archaeologists believe that more than thirty thousand years ago humans carved these shapes into a piece of bone to track the phases of the moon.

Early Predictions

Natural cycles have always been important to life on Earth. Even before humans understood why day turned to night and the seasons changed, early humans organized their lives around these events. In fact, artifacts from the Stone Age show that even tens of thousands of years ago, humans were tracking the phases of the moon as a means to measure time. Some very early written works also concern astronomical cycles. The Venus Tablet of Ammisaduqa (from the civilization of Babylon) is a record of when the planet Venus rose and set for more than twenty years in the first millennium BCE.

During this early period in human history, astronomers were unable to determine the true reason for why these phenomena followed cyclic patterns. The evidence for our current understanding of the universe, with Earth orbiting the sun and the moon orbiting Earth, had not yet been discovered. But their observations and scientific theories were a crucial foundation for our modern understanding.

THE STONE AGE

The earliest artifact that appears to portray the phases of the moon dates to around thirty-four thousand years ago. However, there is debate as to whether this ancient artifact tracks the phases of the moon (and, therefore, lunar months) or not. Some academics claim that the small circular and crescent-shaped markings on shards of bones represent the moon, while other academics do not believe there is sufficient evidence for this claim.

The connection between ancient megalithic monuments (large monuments made from stone), such as Stonehenge, and astronomy is also quite controversial. Some argue that Stonehenge functioned as an astronomical observatory, allowing people thousands of years ago to predict eclipses, while many academics state this view is unsupported by the site. *The Cambridge Illustrated History of Astronomy* sums up the debate in the following way:

> *In short, there is good reason to think that the construction of Stonehenge and related monuments embodied astronomical symbolism, but we have as yet no convincing evidence that what*

we might think of as scientific astronomy was practiced there.

What is *not* controversial is the importance of the heavens to prehistoric people and the earliest civilizations. The sun, moon, and stars were used to track time, for navigation, and as signals for when to plant crops. The seasons and the sun and moon often played an integral part in early human rituals, religions, and myths.

BABYLON

Mesopotamia, located across modern-day Iraq and neighboring countries, was home to many important early civilizations, including Sumer (perhaps the first civilization in the world), the Akkadian Empire, Babylonia, and the Neo-Babylonian Empire. The earliest known writing system, Sumerian cuneiform, was developed in this region. In ancient times, there was no sharp distinction between astronomy (the science of celestial objects) and astrology (the study of celestial objects to predict events, for example, through horoscopes). And, in fact, early attempts at astronomy were sometimes driven by a desire to know when "omens" like eclipses would occur. Nevertheless, astronomers in the Neo-Babylonian Empire made a number of important astronomical discoveries.

They were the first people to realize that eclipses recurred after a fixed number of days: 6,583.3 days (or 18 years and 11.3 days) to be exact. They named this a "Saros" cycle, and this Babylonian word is still used by scientists to this day. Babylonian astronomers also realized that the moon does not move at a constant rate; it accelerates and decelerates depending on its location. Although they could not yet explain this phenomena, it revealed their careful scrutiny of celestial objects.

ANCIENT GREECE

Ancient Greece is often called the "cradle of Western civilization" because of its historical importance. It was there that democracy was born, and it was there that early philosophers rejected myth in favor of reason as an explanation of natural phenomena. Influenced by Babylonian astronomy, Greek philosophers continued to investigate celestial objects. The word "planet" even comes from the Greek word for "wanderer"; this etymology refers to the planets' independent movement compared to the stars in the sky. Additionally, the Greeks created names for the various constellations (or groups of stars) that are visible in the night sky.

Like the earlier Babylonians, the Greeks were fascinated by eclipses. Aristotle, the famous Greek philosopher, concluded that Earth must be a sphere from observing lunar

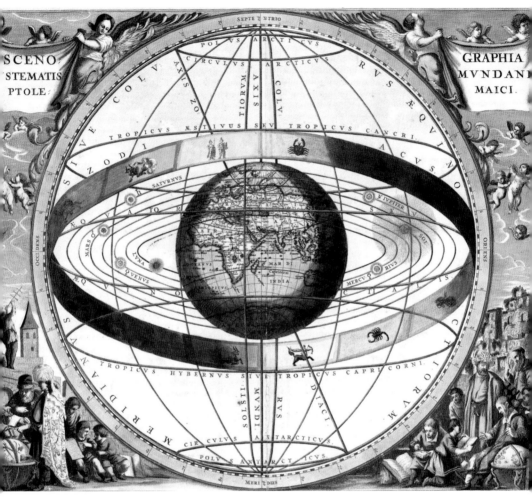

Early models of the solar system placed Earth at the center rather than the sun.

eclipses. In every eclipse, the shadow cast onto the moon from Earth is round, no matter what angle the sun is behind Earth. Therefore, he argued that Earth must be spherical rather than flat. If Earth were a flat disk, the shape of the shadow would differ depending on the angle of the sun.

Early Predictions

Ptolemy, a later Greek thinker, was able to predict eclipses based on data from the past. But, like Aristotle, Ptolemy believed the sun revolved around Earth rather than the opposite. His **geocentric model** of the solar system (also called the Ptolemaic system) placed Earth at the center of the universe. Despite this flaw, the Ptolemaic system would be used for over a thousand years to predict the movement of the celestial bodies with relative precision.

With regards to tides and the structure of the solar system, some Greek thinkers first posited a number of ideas that were later proven to be true, although at the time they were unable to prove their hypotheses. In the fourth century BCE, the Greek geographer and explorer Pytheas made a number of important contributions to science. He was the first person to theorize that tides were caused by the moon, and he also measured the tilt of the Earth. Later, in the third century BCE, Aristarchus of Samos became the first astronomer to argue that Earth revolves around the sun. While his theory was not accepted at the time, his **heliocentric model** of the solar system (with Earth revolving around the sun) would later revolutionize our understanding of the universe.

The Precession of the Equinoxes

Hipparchus was a Greek astronomer who lived in the second century BCE. A meticulous observer of the night

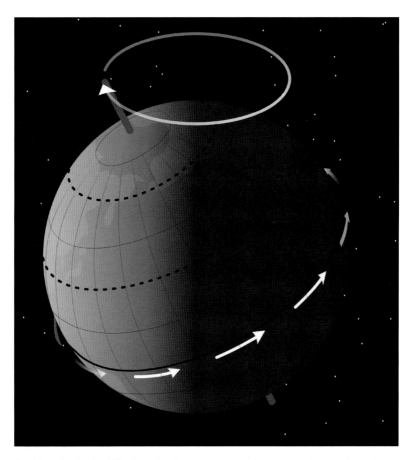

Earth's axis slowly shifts in a circular pattern over the course of approximately twenty-six thousand years.

sky, he made a startling discovery when he compared his own observations to those of Greek astronomers from the previous century: the position of the stars seemed to have shifted in the night sky. On the day of the **equinox** (when day and night are of almost equal length), it was thought the stars should be in the same position as last year. But Hipparchus proved that this was not the case—an important discovery given that astronomers used the stars' position as a fixed frame of reference for their observations. He named this phenomenon the **precession of the equinoxes** because the date of the equinox seemed to precess (or move) when compared to the stars in the sky.

Although Hipparchus did not know it at the time, the precession of the equinoxes is caused by the fact that Earth wobbles slightly as it spins on its axis. This movement is often compared to the wobbling of child's top as it spins. But this "wobble" is at an incredibly slow pace. If you imagine a line extends from the center of Earth into the heavens above, it takes twenty-six thousand years for the line to slowly complete a circle against the backdrop of stars. While this motion does not significantly change Earth's tilt, it does change how the stars appear from Earth.

You may know that the bright star Polaris is now called the North Star, but this was not always the case. At different times in human history, other stars have been the North Star. Pytheas even recorded that there was no true North Star by his calculations—only an empty space in the night sky. The North Star is useful in navigations because it and the South Star are the only two stars in the night sky which do not seem to move. The other stars in the sky appear to revolve around them.

It was Hipparchus who realized that the North Star—the one seemingly fixed point of the night sky—is, in fact, only motionless for a given time. The precession of the equinoxes means that even its identity changes over time. The precession of the equinoxes, and the wobbling motion of Earth that causes it, is just one of the many cyclic patterns visible in the universe.

NICOLAUS COPERNICUS

The structure of our solar system had to be understood before we could arrive at a true understanding of the seasons, eclipses, lunar phases, and tides. But with the exception of

During a lunar eclipse, the moon appears to turn red. This ominous sight often terrified early civilizations.

Early Predictions

Aristarchus in Ancient Greece, most scientists did not believe Earth orbited the sun until the groundbreaking work of Nicolaus Copernicus. In 1543, Copernicus published his work *On the Revolutions of the Heavenly Spheres.* He argued that it was the sun at the center of the universe rather than Earth, and he used mathematics and astronomical observations to support this model of the universe. Only after this discovery was it possible for a complete understanding of the causes of the seasons, eclipses, lunar phases, and tides to take form.

Yet Copernicus's theory was not universally accepted by other scientists of his day. A number of flaws meant that his model of the universe was no better at predicting astronomical events than the geocentric model of Ptolemy. Given this imprecision, other astronomers were unconvinced that his heliocentric model was correct because it implied two facts they found implausible: Earth is hurtling through space at an incredibly fast speed, even though it does not appear to be moving from our perspective. Seondly, the stars must be similar in size to the sun and incredibly distant from our own solar system because they do not seem to move as Earth orbits the sun.

While both of these facts are true, many of Copernicus's peers found them more difficult to accept than the time-honored geocentric model. It was not until the works of later scientists that the heliocentric model would become universally accepted, but the works of Copernicus began the

Copernican Revolution that would remove Earth—and humanity—from the center of the universe.

JOHANNES KEPLER

The first key supporter of Copernicus was Johannes Kepler. Although Copernicus correctly theorized that Earth was the center of the solar system, his version of the heliocentric

Johannes Kepler (1571-1630)

Christopher Columbus and an Angry God

In 1502, Christopher Columbus was stranded on the island of Jamaica. His fourth transatlantic voyage had been plagued by problems, and he was forced to rely on the native inhabitants of Jamaica for food while he waited for help from Spain. At first, the indigenous people traded food for trinkets that the sailors owned; however, after Columbus's men plundered several villages, they cut off the explorers' supply. Desperate to acquire food and armed only with an almanac that predicted lunar eclipses, Columbus decided to trick the local people. According to John Rao at http://www.space.com:

> [Columbus] asked for a meeting with the natives' Cacique ("chief") and announced to him that his Christian god was angry with his people for no longer supplying Columbus and his men with food. Therefore, he was about to provide a clear sign of his displeasure: Three nights hence, he would all but obliterate the rising full moon, making it appear "inflamed with wrath," which would signify the evils that would soon be inflicted upon all of them.

When the lunar eclipse did occur, the initially skeptical chief asked Columbus to intercede with his god on his people's behalf. After timing when the eclipse would end, Columbus agreed that his god would end the eclipse if his sailors were given food. The eclipse soon ended and the chief, believing Columbus was responsible, provided his men with food until a ship from Spain rescued them. This dramatic story shows the usefulness of astronomical knowledge. Writers have retold versions of this story since Columbus's famous deception.

model still had some problematic features. It did not predict astronomical events with any greater accuracy than the previous geocentric model. This is because the Copernican system still had incorrect features of the earlier geocentric models, notably the circular orbits of the planets and **epicycles**. Epicycles were hypothesized small circles the planets moved in as they followed a larger orbit. Epicycles were necessary to explain the movement of the planets according to both the geocentric and early heliocentric model of Copernicus. It was Kepler who realized—to his surprise—that the Copernican model could not be accurate when he examined the more precise measurements of the movement of celestial bodies available to him. As a result, he concluded that the planets must follow an elliptical orbit around the sun. This realization simplified the earlier models of planets. Planets went from making a large circular orbit with numerous epicycles along the way to making one elliptical orbit.

Kepler also realized that the speed of the planets must vary depending on their position. This was necessary to explain detailed observations of planetary positions, and it was an important scientific discovery. However, Kepler was troubled by the fact that there was no apparent connection between the various orbits and speeds of the planets: each planet had its own unique orbit and speeds that appeared unrelated to the others. Kepler devised his own explanation for why the planets move in the way that they do, but it was not until Isaac Newton developed his theory of gravity that there was an adequate explanation for the planets' elliptical orbits and varying speeds.

GALILEO GALILEI

Galileo Galilei was another key figure in the Copernican Revolution. A contemporary of Johannes Kepler, he too defended the heliocentric model of the universe when most other scientists of his day doubted it. His major astronomical achievements were based on his use of the telescope. Although he did not invent the telescope, he improved upon early designs, and he was one of the first people to use it to observe celestial objects. He used these observations to defend the Copernican model of the universe. One of the most important discoveries he made was that the planet Jupiter has four moons, although he called them "stars." This was a shocking discovery at the time because it contradicted the geocentric model that said *all* celestial objects orbited Earth. It was such a significant blow against the geocentric model that some of his contemporaries rejected the very idea that the moons of Jupiter existed—ascribing their observation to a defect in telescopes rather than admitting celestial objects existed that did not orbit Earth.

Despite his importance to the Copernican Revolution (and the larger scientific revolution of the time), Galileo held a number of incorrect views about the model of the solar system. He rejected Kepler's correct argument (an argument based on empirical evidence) that the tides were caused by the moon. Galileo attempted to use the tides to prove the heliocentric model, arguing that the tides were caused by Earth's movement around the sun (the orbit causing the water on Earth to slosh about) rather than the pull of the moon. This turned out to be incorrect, as Isaac Newton was soon to prove with his theory of gravity.

The phases of the moon take place over the course of 29 days, 12 hours, 44 minutes, and 3 seconds on average.

The Modern Understanding of Lunar Phases, Eclipses, and Seasons

In order to understand the cycles that result from the movement of the sun, Earth, and moon, it is necessary to understand why these objects move the way that they do. We have already looked at the evolution of human understanding from the geocentric model to the heliocentric model and from circular orbits to elliptical orbits. Now, we can examine why the planets move in elliptical orbits around the sun (although the planets actually orbit the center of mass of the solar system, which is a changing point *next* to the center of the sun).

GRAVITY

After the astonishing discoveries of Copernicus, Kepler, and Galileo, Isaac Newton cemented the heliocentric model of the universe by explaining its cause: gravity. His

Isaac Newton (1643-1727)

work, considered among the most important in the field of science, would once and for all silence serious doubts about the heliocentric model of the universe. It would also do what Kepler had tried and failed to do: provide a unifying explanation for why the planets move at the speed that they do.

On July 5, 1687, Newton published his *Philosophiæ Naturalis Principia Mathematica*, often called the *Principia*. The importance of this work in the history of science is monumental.

Newton's *Principia* established classical mechanics (also called Newtonian mechanics), the set of scientific laws that describe how objects move. The most important law for the purposes of this book is the one that primarily governs the movement of the celestial objects: gravity. Newton was the first to discover that the same force that causes objects to fall on Earth causes Earth and the moon to orbit. Additionally, he realized that every particle in the universe attracts every other particle in the universe to a degree that is determined by their mass and distance from one another. His law of universal gravitation can be represented as the following equation in the language of today:

$$F = G \frac{m_1 \times m_2}{r^2}$$

F is the force (of gravity), G is the gravitational constant, m_1 is the first mass, m_2 is the second mass, and r is the distance between the two objects. As we can see from this equation, the force is stronger the greater the mass of the objects, and it is weaker the greater the distance between the two objects.

While the equation representing Newton's law of universal gravitation is simple in its modern format, it is quite difficult to use it to calculate the movement of the celestial objects given the complex way the planets interact with one another, their moons, and the sun as they orbit. Some scientists were initially skeptical of Newton's claims, but their doubts were soon laid to rest. Edmond Halley used Newton's work to predict when a comet (now called Halley's Comet) would be visible given the effects of Jupiter and Saturn's gravity on the comet's orbit around the sun. This was the first time any object other than a planet was shown to orbit the sun, and Halley's ability to precisely predict its return was evidence of the accuracy of Newton's laws.

A later discovery would further prove the usefulness of Newton's laws. As it is described in *The Cambridge Illustrated History of Astronomy*:

> *The discovery of Neptune in 1846 was the ultimate triumph of Newtonian dynamics: two astronomer-mathematicians, sitting at their desk, had calculated from the effects—the deviations of Uranus from its predicted orbit—to the cause, and had pinpointed the whereabouts of the culprit, a major planet whose very existence had until then been unsuspected.*

This kind of precision that enabled a planet to be discovered via mathematics was unprecedented in astronomical models before Newton.

Newton's theory of universal gravitation revolutionized our understanding of the heavens. It explained the existence of the tides, the reason for Kepler's elliptical orbits, and why the moon and planets sometimes deviated from their simple orbits (due to the gravitational force exerted by nearby planets). To this day, Newton's laws are still used in astronomy to predict the approximate movement of celestial objects, though Albert Einstein's general theory of relativity has replaced Newtonian mechanics as a more precise description of gravity. We will examine the differences between Einstein's and Newton's theories later in this chapter.

THE SEASONS

One common misconception is that the seasons are caused by Earth's proximity (closeness) to the sun. This is not true. While the distance between Earth and the sun does vary depending on the time of year, it does not significantly impact the temperature on Earth. In fact, Earth is closest to the sun in January when it is winter in the United States. The seasons are actually caused by Earth's tilt. But what exactly do we mean when we say Earth is tilted? Tilted in relation to what? To answer this question, we need to remember that

When the Northern or Southern Hemisphere is tilted towards the sun, it is summer there; when tilted away, it is winter.

Earth is following an orbit (an elliptical path determined by gravity) around the sun. This orbit lies in an imaginary plane (a two-dimensional surface) that is called the orbital plane. Earth's axis is tilted relative to this orbital plane. In other words, Earth's equator does not line up with Earth's orbit.

This **axial tilt** means that the two hemispheres (Northern and Southern) of Earth receive different amounts of sunshine depending on where Earth is along its orbit. The Northern Hemisphere is the half of Earth that is north of the equator while the Southern Hemisphere is the half south of the equator. At any point in time, the Northern and Southern

Hemispheres of Earth experience opposite seasons. For example, when it is summer in the Northern Hemisphere, it is winter in the Southern Hemisphere. This is because the hemisphere of Earth that is tilted towards the sun receives more sunlight. This results in more hours of daylight and higher levels of **solar radiation** (energy from the sun) during the summer.

There are two different ways of defining the four seasons: meteorologically and astronomically. The astronomical seasons are based on the length of the day, while the meteorological seasons align the summer and winter to the three hottest and coldest months of the year. Thus, summer in the Northern Hemisphere is June, July, and August while those three months are winter in the Southern Hemisphere. Spring and fall naturally occupy the three months between summer and winter. However, most countries around the world, including the United States, follow the astronomical seasons rather than the meteorological seasons.

The astronomical seasons are aligned with the equinoxes and **solstices**. There are two equinoxes and two solstices every year. The summer solstice is the longest day of the year, and the winter solstice is the shortest day of the year. The terms "June solstice" and "December solstice" are also used because they always occur during these months of the **Gregorian calendar**. However, these terms leave it ambiguous as to whether the solstice described is the shortest or longest day of the year because it depends on the hemisphere from which the solstice is observed.

The vernal and autumnal equinoxes take place on the day that the equator crosses the center of the sun. Many people think that this means day and night are the same length on this day (twelve hours), but this is actually not the case. On the day of the equinox, the geometric center of the sun is above the horizon for twelve hours and below the horizon for twelve hours. However, the sun appears to be a large disc from Earth, not a two-dimensional point. Therefore, at sunrise the edge of the sun is actually visible above the horizon for a short span of time before the geometric center is above the horizon, and at sunset the edge of the sun is once again visible after the geometric center of the sun is below the horizon. This means day is longer than night on the day of the equinox.

According to the astronomical reckoning, winter begins on the winter solstice, spring begins on the vernal equinox, summer begins on the summer solstice, and fall begins on the autumnal equinox. In the Gregorian calendar, the equinoxes and solstices fall back by almost precisely six hours each calendar year for four years until they are reset by the leap year. In this system, summer then starts on the longest day of the year. While you may think this would be the hottest day of the year on average and should, therefore, be the middle of summer, this is not the case because of **seasonal lag**. Seasonal lag refers to the fact that changes in Earth's temperature actually lag behind the changes in sunlight that cause them. This is because it takes time for Earth to heat up and cool off as a result of increasing or decreasing sunlight.

Another curious fact about the astronomical seasons is that they differ from one another in length. Currently, winter is eighty-nine days, spring is ninety-three days, summer is ninety-three days, and fall is ninety days. These values are approximations and constantly changing. These differences in the lengths of seasons are due to the elliptical orbit of Earth and the effect of gravity. When Earth is closest to the sun, it moves at a greater rate of speed than when it is farther from the sun. Along the present orbit of Earth, this occurs during the winter, although this will slowly change over time. From this, we can see that the seasons and their lengths are directly caused by the effects of gravity and Earth's elliptical orbit.

THE CALENDAR

Today, we use the Gregorian calendar, and to the casual observer, very little appears to change from year to year. The seasons remain roughly the same (never varying by more than a couple days) and the days of the year seem to fit neatly into the calendar with the minor imperfection of a leap year every four years. In fact, the Gregorian calendar resulted from thousands of years of trial and error and scientific reasoning. The rules of the calendar are more complicated than most people know, and the difficulty of creating a calendar that

Julius Caesar, famous for his conquests and his appearance in a play by Shakespeare, also played an important role in the development of our calendar.

does not create inaccuracies over time is one that dogged scientists for thousands of years.

The crux of the problem is that a solar year is approximately 365.242 days instead of a whole number. This is unsurprising considering the fact that the length of the day is determined by the time it takes Earth to rotate on its axis, and the length of the year is determined by the time it takes Earth to orbit the sun—they are not logically related durations of time even though we often think of them together. Before our current Gregorian calendar, the Julian calendar (named after the man who instituted it, Julius Caesar) was used throughout much of the Western world. Prior to the adoption of the Julian calendar, the calendar year had drifted so much that Julius Caesar was forced to decree that 46 BCE would have 445 days in it to bring the months back into alignment with the solar year. The Julian calendar is nearly identical to the Gregorian calendar of today with one crucial difference: there was a leap year *every* four years (unlike today when there are exceptions to the four-year rule as described below). As a result, the Julian calendar would not drift if the solar year were 365.25 days, but this still does not precisely align with the real value of 365.242.

In 1582, Pope Gregory instituted the Gregorian calendar in order to reduce this inaccuracy. He was chiefly concerned with the holiday of Easter. Easter had initially been aligned to the vernal equinox in 225 CE; however, by the year

The Eclipse of 1919

One implication of Albert Einstein's theory of general relativity was that gravity would affect light as well as objects like the planets. This differed from Newton's laws and, therefore, the measurement of gravity's effect on light would provide important evidence for or against the theory of general relativity. In 1911, Einstein wrote the following passage about the importance of this topic:

> *In a memoir published four years ago, I tried to answer the question whether the propagation of light is influenced by gravitation. I return to this theme, because my previous presentation of the subject does not satisfy me, and for a stronger reason, because I now see that one of the most important consequences of my former treatment is capable of being tested experimentally. For it follows from the theory here to be brought forward, that rays of light, passing close to the sun, are deflected by its gravitational field, so that the angular distance between the sun and a fixed star appearing near to it is apparently increased by nearly a second of arc.*

The only problem was the bright light of the sun made it impossible to do such an experiment. But in 1919, Arthur Eddington was able to complete the experiment during a total solar eclipse, when the light of the sun was blocked by the moon and a particularly bright star was nearby. The results matched Einstein's prediction, providing early evidence for the theory of general relativity and making Einstein an international celebrity overnight.

1582 CE, Easter fell ten days after the vernal equinox. This happened because the Julian calendar was gaining eleven minutes a year compared to the actual solar year. Pope Gregory modified the occurrence of leap years to fix this problem. According to the Gregorian calendar, every year divisible by four is a leap year unless it is divisible by one hundred—with the further exception that years divisible by four hundred *are* leap years. This minor change significantly reduced the rate at which the calendar year drifted out of alignment with the solar year. As we can see, the apparently simple calendar we use is quite complex in order to account for the fact that the solar year does not contain a whole number of days.

LUNAR PHASES

The phases of the moon are caused by the changing positions of the sun, Earth, and moon in relation to one another. Roughly half of the moon is always illuminated by light from the sun. When this half of the moon is facing Earth, it is known as a full moon, and the moon appears to be a bright disc in the sky. When the moon is full, the moon is on the opposite side of Earth as the Sun. Therefore, light from the sun illuminates the side of the moon that is currently facing Earth. However, when the moon is on the same side of Earth as the sun, the sun illuminates the half of the moon that is not visible from Earth. We call this a new moon, and the moon appears completely dark in the sky, although it is still barely visible in the sky because of reflected light from Earth (called earthshine).

There are four principal lunar phases: new moon, first quarter, full moon, and third quarter. During a new moon, the entire visible area of the moon is dark; during a first quarter moon, the right half of the moon is illuminated; during a full moon, the entire visible area of the moon is illuminated; and during a third quarter moon, the left half of the moon is illuminated. After progressing through these four phases, the lunar phases then repeat once again after the third quarter moon becomes a new moon. Confusingly, the first quarter and third quarter moons are also referred to in casual speech as a "half moon" because half of the visible area of the moon is illuminated. In addition to the four principal phases of the moon that only last one night each, there are intermediate phases as well. These four intermediate phases are named by adding the word "waxing" (if the illumined area is increasing) or "waning" (if it is decreasing) to the word "crescent" (if the illumined portion is less than half) or "gibbous" (if it is more than half). For instance, after the new moon, the first intermediate phase is the waxing crescent because the illumined area of the moon is growing but less than half is illuminated.

Interestingly, the area of the moon that is visible from Earth, whether illumined by the sun or dark except for earthshine, is always the same portion of the moon. This is because the moon is **tidally locked** to Earth—meaning the moon takes precisely the same amount of time to rotate around its axis as it does to orbit Earth one time. We always see the same side of the moon, while Earth is spinning when looked at from the moon. This phenomenon is quite common among planets and their satellites. In one notable case, Pluto and its moon Charon are both tidally locked to one another, meaning an observer on either object would only be able to see one side of the other.

It takes 27.3 days for the moon to orbit around Earth—this is called a sidereal month. You might think the lunar phases would align with this number, but they do not. This is because while the moon is orbiting Earth, Earth is also orbiting the sun. In those 27.3 days, Earth has progressed nearly a twelfth of the way around the sun. Since the Earth's position relative to the sun has changed, the moon must catch up to return to the same position relative to the sun. This adds 2.2 days to the time it takes the lunar phases to occur. As a result, the lunar phases take place over a period of 29.5 days—this is called a synodic month. While the lunar phases are the historical reason for the months of the year, we now use a solar calendar, not a lunar calendar, and the twelve months of the Gregorian calendar are not connected to the phases of the moon.

ECLIPSES

An eclipse occurs when a celestial body or its shadow obscures the light from another celestial body. This can occur when one object passes in front of another or when an object blocks light from reaching another object. On Earth, the most noticeable eclipses are solar eclipses (when the moon blocks light from the sun from reaching Earth) and lunar eclipses (when Earth blocks light from the sun, preventing it from illuminating the moon). Eclipses can occur only when the sun, moon, and Earth are roughly in a straight line: solar eclipses can only take place around a new moon, while lunar eclipses can only take place around a full moon due to the relative positions of the sun, Earth, and moon. But eclipses do not happen every new and full moon because the orbit of the moon is tilted relative to the orbit of Earth around the sun. As a result, there are only two eclipse seasons each year. The eclipse seasons last roughly thirty-five days, and at least one solar and lunar eclipse always occur, although solar eclipses are only visible from a small area of the Earth's surface, and they may not be total eclipses.

The most dramatic type of eclipse is the total solar eclipse, when the moon completely covers the sun and the sky darkens. But this only occurs on a very small area of the Earth during an eclipse because of the size and position of the moon—the moon's shadow is not big enough to cover the entire Earth. A total solar eclipse is also only possible at certain points in the moon's elliptical orbit (when the moon is

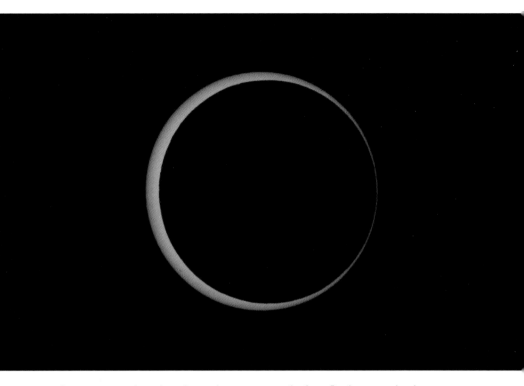

During an annular solar eclipse, the moon is too far from Earth to completely block the sun from sight.

closest to Earth). When the moon is farther from Earth, the apparent size of the moon is smaller than the apparent size of the sun. This is called annular eclipse, and the outer edge of the sun is visible around the black void of the moon. In this way, the elliptical orbit of the moon results in two kinds of eclipses that look radically different. The most common kind of solar eclipse is a partial eclipse when the moon and sun are not perfectly aligned, so the moon only covers a portion of the sun. Usually, these are not even noticeable because the sun does not appear to dim significantly.

Lunar eclipses can also be partial or total depending on whether Earth's shadow covers the entire surface of the moon. Unlike solar eclipses, which are only visible in a limited area, lunar eclipses are visible from anywhere the moon is visible. This is because Earth is much larger than the moon, and its shadow obscures the entire moon, while the shadow of the moon only falls on a small part of Earth. As a result, lunar eclipses also last for hours while solar eclipses last for just minutes. The moon appears to slowly darken and take on a reddish hue during a lunar eclipse. This reddish color is due to the refraction of sunlight through clouds and dust of Earth's atmosphere (sunsets and sunrises appear red for the same reason). Because of this, total lunar eclipses are sometimes referred to as blood moons.

A NEW CONCEPTION OF GRAVITY

The seasons, lunar phases, and eclipses can all be explained according to Newton's law of universal gravitation and the model of the universe based on this law. But this is not how scientists now believe the universe works. Mercury's orbit was one early sign of the shortcomings of Newton's laws of motion. Astronomers quickly realized that its orbit could not be explained by Newton's theory of universal gravitation. The actual orbit differed by a few seconds from the orbit calculated using Newton's laws. At first, it was theorized a small planet, named Vulcan, existed even closer to the sun than Mercury, and its gravitational field was affecting Mercury's orbit; however, astronomers were unable to locate Vulcan, and it is no longer thought to exist. The mystery of Mercury's orbit would not be solved until Albert Einstein set forth his theory of general relativity.

In Newton's mind, time passed uniformly and space was the vast field in which objects moved. But we now know this is not the case. The currently accepted theory, the theory of general relativity, says that space and time form a continuum. This continuum is curved by the presence of matter. The bending of **space-time** that results from the sun causes the elliptical orbits of the planets. According to this conception of the universe, gravity is not an independent force that pulls objects. Instead, it is a result of the curvature of space-time. Einstein's equations were able to explain the orbit of Mercury more precisely than Newton's laws, but scientists at the time were initially skeptical of Einstein's ideas. His theory of general relativity was not widely accepted until a number of experiments confirmed predictions that Einstein had made.

Einstein's theory of relativity did more than explain why gravity causes objects to move. It also led to a new understanding of the universe. Time is relative according to Einstein, rather than absolute as it was to Newton. The same event can occur at two different times to two different observers. The theory of general relativity also explained the existence of black holes and other astronomical discoveries.

This statue of the famous Greek explorer Pytheas is in his hometown of Marseilles—now part of France.

3

Scientists and Mathematicians

Countless astronomers and scientists contributed to our current understanding of Earth, the sun, and the moon. We will look at the lives and achievements of just a few of the most important. Beginning with Kidinnu in the fourth century BCE and ending with Albert Einstein in the twentieth century, these diverse individuals struggled with different individual and political challenges while completing their important work.

KIDINNU

The Babylonians were among the earliest people to record astronomical events in writing. These records later proved to be quite useful for modern scholars who used them to establish when ancient events occurred. Archaeologists have found clay tablets that the Babylonians recorded their astronomical observations on, and some of their astronomical

The Babylonians wrote their astronomical observations on clay tablets. This is one that records the movements of Venus.

ideas and theories have reached us through the writings of ancient Greek and Roman authors. But our understanding of their work is still incomplete, and we know very little about the individual astronomers who dedicated their lives to understanding the heavens.

One Babylonian astronomer whose name did reach us was Kidinnu. He almost certainly lived during the fourth century BCE, although it is not known when he was born. He is credited with one of the great achievements of Babylonian astronomy: System B. System B is a way of reckoning the position of the moon and the lunar phases with a high degree of precision. It was developed because the Babylonians realized that the moon does not always move at the same speed—it accelerates and decelerates (we now know this is because of its elliptical orbit). System B links the speed of the moon to the lunar phases and also provides calculations to accurately predict the moon's future movements. This was a groundbreaking discovery for the time.

Kidinnu is also sometimes credited with reforming the Babylonian calendar. As we saw in chapter 2, the fact that the solar year is not made up of a whole number of days greatly complicates the formulation of a precise calendar. But an even earlier complication was the fact that a lunar month has approximately 29.5 days. Twelve lunar months therefore equal 354 days—roughly 11¼ days short of a solar year. Like the Romans and Julius Caesar in later times, the Babylonians added leap months to correct their calendar as needed. Sometime in the sixth century BCE, the Babylonians realized that there are almost exactly 235 lunar months in 19 solar years. This meant a system could be developed where leap months would be determined ahead of time on a 19-year cycle. The lunisolar Hebrew calendar, used to determine Jewish holidays, still uses this 19-year schedule of added leap months. However, the 19-year cycle proved to be imprecise over large stretches of time, and the calendar was reformed in the fourth century BCE, possibly by Kidinnu. The time period was extended from 19 years to 76. This new system was so exact that after Alexander the Great conquered the city of Babylon, the calendar was sent back to Greece and immediately adopted there.

Very little is known of Kidinnu's life, although it is likely he was a priest because priests were responsible for recording astronomical observations and making predictions about the future based upon their findings. The exact years he lived are unknown, but a clue may be found in a series of fragmentary clay tablets that record the events after Alexander conquered the city of Babylon in 331 BCE. It states simply, "Kidinnu was killed by the sword." While there is no definitive proof that this person named Kidinnu was the same famous astronomer whom later Greek and Roman authors reference, archaeologist Jona Ledering argues:

This Kidinnu must have been someone well-known to the author, because he is mentioned without any familial or professional designation. As the Babylonian chronicles were written by the same scribes as the Astronomical diaries and other astronomical texts, it is tempting to think that the astronomer fell victim to Alexander's enlightened science policy. If this identification is correct, the inventor of System-B must have been an old man when he was executed.

This rare evidence of an executed Babylonian man's name gives a human face to ancient people who lived so long ago and who first began to shape our understanding of the heavens.

PYTHEAS

Pytheas was a Greek explorer and geographer who lived soon after the time of Kidinnu. While historians do not know precisely when either man lived, it is probable that Pytheas was born while Kidinnu was an old man or had recently died. Pytheas was born in the Greek colony of Massalia, which is now the modern city of Marseille in southern France. Marseille is now nearly one thousand miles from Greece, but at the time Greek colonies formed a ring around much of the Mediterranean Sea. Overpopulated Greek city-states would send expeditions to establish new cities along the coast, and young Greeks would seek opportunities there. Massalia was founded by colonists from Phocaea (a Greek city which was located in modern-day Turkey) a couple hundred years before the birth of Pytheas in the middle of the fourth century BCE.

Pytheas is most famous for a daring voyage he made from the Mediterranean Sea, up and around the British Isles, and on to an even more distant land (perhaps Iceland or Norway). Pytheas's account was the only written record of the British Isles until Julius Caesar launched his invasions of Britain three hundred years later. Pytheas also ventured inland and described the lives of the indigenous tribes that lived there.

In the ancient world, many authors were skeptical of Pytheas's claims, but today many historians think his accounts are accurate. In particular, his descriptions of the sea near the Arctic Circle as a mixture of sea, earth, and vapor that was impossible to walk or sail on drew ridicule. But now many historians think this was a description of drift ice—pieces of ice that sometimes form a near-continuous sheet on the ocean and can be accompanied by fog.

In addition to his remarkable voyage, Pytheas made a number of scientific discoveries. His writings, like those of many ancient Greek and Roman authors, did not survive, but we know some of what he wrote because other authors cited and discussed his works. Pytheas was the first person we know of who realized the tides were influenced by the moon. This was possible because of his voyage outside of Mediterranean Sea. Tides in the virtually landlocked Mediterranean average only a couple inches, making scientific observation difficult. But in the open waters around Britain the tides are much easier to observe—in the Severn Estuary between England and Wales, for instance, the water level can vary by 50 feet (15.24 meters) depending on the tide.

Pytheas also made a number of discoveries by measuring the shadow of a gnomon—the part of a sundial that casts a shadow. His measurements led him to determine the axial tilt of the earth with a great deal of precision. This was a new discovery at the time and proved valuable for subsequent astronomers. He could also accurately calculate the latitude (the distance north or south of equator) of various places. This proved a valuable skill on his voyage because it allowed him to make detailed records about his location.

GALILEO GALILEI

Galileo Galilei was born in Pisa, Italy, in 1564—twenty-one years after Copernicus first published his arguments for the sun being the center of the universe. He would become one of the most important supporters of the heliocentric model, and his personal life would be greatly affected by his scientific pursuits. Having originally entered the University of Pisa to study medicine, he later changed his course of study to

Galileo worked on his scientific theories for many years under house arrest.

mathematics. He made numerous important contributions to science in his lifetime, and he was an important figure in the scientific revolution of the time. Professor Peter Machamer sums up his importance by saying, "When he was born there was no such thing as 'science,' yet by the time he died science was well on its way to becoming a discipline and its concepts and method a whole philosophical system."

Many of Galileo's astronomical discoveries were made possible by his use of the telescope. He first heard about the invention of the telescope in 1609, and months later he had built a functional model of his own. His first model, like the ones built by the Dutch inventors, only magnified objects to appear three times closer than the naked eye. But he soon improved upon this early design and developed a telescope that made objects appear twenty times closer. This led to an astonishing number of important astronomical observations. According to Professor Noel M. Swerdlow, "In about two months, December and January, he made more discoveries that changed the world than anyone has ever made before or since. He began with the irregular surface of the moon, went on to the uncountable number of stars, and then in early January found the early satellites of Jupiter."

Another important discovery Galileo made with the telescope was that the planet Venus had phases like the moon. Sometimes it appears to be a (nearly) circular disc, while other times it appears to be a crescent. This was extremely damaging evidence to the geocentric model of the universe: scientists of the era were aware that the phases of the moon were caused by light from the sun illuminating the moon, but the same thing could not happen to Venus if it were orbiting the Earth and closer than the sun—the light from the sun would always fall primarily on the side of Venus that was not visible from the Earth.

Many people know about Galileo because of his public conflict with the Catholic Church—now known as the Galileo affair. At the time, the Inquisition was underway in Europe. Institutions of the Catholic Church were combating heresy (ideas deemed to be out of line with church doctrine) by banning or censoring books, compelling scientists and theologians to recant unorthodox views, and sometimes imprisoning or executing those found guilty of heresy. The Protestant Reformation that began in the early 1500s had split the Catholic Church, and the Inquisition had reached a feverish intensity in Italy—the Catholic stronghold and seat of the papacy—as a result.

The Catholic Church had remained silent about the Copernican model of the universe, which was first published in 1543; Copernicus's book was even dedicated to Pope Paul III. But in 1616, this changed. The Inquisition decreed that the heliocentric model was "contrary to Scripture" and ordered Galileo, a well-known supporter of the theory, to stop defending the heliocentric model. Copernicus's work was also "corrected" so that it no longer promoted the heliocentric model. This settled the matter until 1632, when Galileo published another work that concerned the heliocentric model. Galileo assumed that his close relationship with the Pope and his evenhanded treatment of the topic (the book also presented arguments against the heliocentric model) would shield him from the Inquisition, but he was summoned to Rome and famously condemned for being "vehemently suspect of heresy." He was forced to live under house arrest for the rest of his life and also forced to publicly reject the heliocentric model. At his home, he continued to work on scientific ideas for the next nine years, until his death in 1642.

Galileo was never married, but he did have three children with the same woman, Marina Gamba. His son was a lutenist (a lute is a string instrument that was popular at the time) like Galileo's father, and his two daughters were nuns. Originally buried in a modest tomb because of his condemnation by the Church, his body was later moved to an ornate tomb in the Basilica of Santa Croce in Florence, Italy, and this is his current resting place.

ISAAC NEWTON

The Copernican model, defended by Galileo and Kepler, would remain controversial until Isaac Newton published his work regarding gravity. Isaac Newton was born in 1643 (or 1642, according to the Julian calendar that England still

The Einstein Museum

Despite his importance and fame, there are relatively few museums that devote space to Albert Einstein year round. In fact, the only permanent exhibition in the United States is housed in a section of a retail store in Princeton, New Jersey, next to Princeton University. Various objects related to Einstein, and especially his time in Princeton, are on display for visitors to look at. But there is a more extensive permanent exhibit at the Bernisches Historisches Museum in Bern, Switzerland—a place Einstein lived when he was developing the theory of relativity. The museum brochure states the following about the genesis and composition of the exhibit:

> 350,000 visitors from all over the world saw the Jubilee Exhibition on the life and work of this genius of physics in 2005/06. Reason enough for the Bernisches Historisches Museum to present the show in concentrated form as a permanent exhibit under the name 'Einstein Museum.' In an area of 1000 square meters, lavishly staged original memorabilia, written records, and film documentaries describe Einstein's life and at the same time illustrate the history of the twentieth century. Animation films, experiments and a virtual journey through the cosmos explain Einstein's revolutionary theories in a clear and easy-to-understand way.

used then), around the time of Galileo's death and nearly one hundred years after Copernicus first published his heliocentric model. The son of a yeoman—a social standing above the lower class but below the gentry—Newton attended Trinity College, Cambridge, after a short-lived career as a farmer.

At the beginning of his academic career, Newton was an unremarkable student. But when the University of Cambridge closed for two years due to an outbreak of the plague, Newton devoted his time at home to study and research. It was at this time that he began developing calculus—the mathematical study of change. Newton's development of calculus would lead him into a bitter dispute with the German mathematician and philosopher Gottfried Leibniz. Both Newton and Leibniz are now credited with developing calculus independently (with different mathematical notations), but during their lives, there was a great controversy over who first developed it. It is now accepted that Newton developed it first, though he did not publish his findings for many years. This allowed Leibniz to publish on the subject before Newton. Today, the mathematical notation of calculus is much more closely related to the notation of Leibniz than that of Newton.

Newton is widely considered one of the most important scientists—if not one of the most important people—in history. His name, like that of Einstein, is synonymous with scientific achievement and objective reason. But it came to light in the 1940s, centuries after his death, that Newton was also interested in the occult. He devoted a great deal of his time to the study of alchemy (trying to find a way to transmute other metals into gold and create an elixir of immortality). His work on this subject was less than scientific. Biographer Michael White characterizes one of Newton's unpublished alchemical books as "little more than a blend of naked delirium and false conviction—the work of a man on the edge of madness."

It was the famed economist John Maynard Keynes who first broke the story of Newton's obsession with alchemy after buying and reading many of his unpublished papers. In 1942, Keynes gave a lecture summarizing his own opinion on the legacy of Newton:

> *In the eighteenth century and since, Newton came to be thought of as the first and greatest of the modern age of scientists, a rationalist, one who taught us to think on the lines of cold and untinctured reason. I do not see him in this light … Newton was not the first of the age of reason. He was the last of the magicians, the last of the Babylonians and Sumerians, the last great mind which looked out on the visible and intellectual world with the same eyes as those who began to*

> *build our intellectual inheritance rather less than 10,000 years ago.*

Whether you regard Newton as a great scientist or as the "last of the magicians," his intellectual work remains his greatest achievement. He dedicated his life to better understanding the world at the expense of his personal life. He never married or had a family, preferring to spend his time pursuing his theories and experiments.

PIERRE-SIMON LAPLACE

Pierre-Simon, marquis de Laplace, was born in 1749, over two decades after Isaac Newton's death. He would build upon the work of Newton and make important contributions to the fields of mathematics, astronomy, and physics. His family was fairly well off, but they were not part of the French upper class, and Laplace initially planned on becoming a priest until he attended mathematics lectures while studying theology at Caen University. Despite these relatively humble beginnings, his aptitude for mathematics would lead him to rub elbows with Napoleon Bonaparte and become a prominent European intellectual.

Laplace made numerous breakthroughs in diverse fields of science and mathematics. One of his major astronomical breakthroughs regarded the **stability of the solar system**.

Pierre-Simon Laplace

According to early geocentric and heliocentric models, the solar system was stable and the orbits of celestial objects were fixed. But Newton's theory of universal gravitation swept away this idea—planets orbited the sun, but their orbits were affected by the gravity of nearby planets as well. This raised a big question: are the orbits of the planets constantly changing (and, therefore, chaotic) or are they stable? Newton himself thought divine intervention must be required to keep the entire system stable. However, Laplace was able to demonstrate that small changes in the orbits of planets were only temporary oscillations and not evidence of long-term variations in the orbits of planets. Scientists now know the solar system is in fact chaotic, and not stable—the orbits of planets cannot be determined beyond tens of millions of years into the future—but Laplace's evidence that the solar system was relatively stable over smaller time frames was important in his day.

Some of Laplace's most enduring scientific work concerned tides. Newton had accurately described what caused tides: the gravitational attraction of the moon (and sun) on the Earth's oceans. This had long been suspected by previous thinkers, like Pytheas and Kepler, due to observational evidence about the cyclical nature of the tides and similar cycles of the moon. Laplace expanded on Newton's work by taking into account a variety of factors, such as the inertia caused by the Earth's rotation, and created an equation to describe the dynamics of water in the oceans. Laplace's equation is still used in the present day, and Laplace's contribution to our understanding of the tides is so great that the book *Tides in Astronomy and Astrophysics* states, "Laplace can be considered as the true founder of the modern science of ocean tides."

Laplace lived during the French Revolution, which began in 1789 and ended in 1799 with the rule of Napoleon Bonaparte. During these years, Laplace was not involved in French politics; some French intellectuals who were politically active lost their lives to the guillotine during the infamous Reign of Terror. But after the ascension of Napoleon—a former student of Laplace's— in 1799, Laplace was made Minister of the Interior for a brief time before being made a senator instead. He was active politically throughout Napoleon's rule, but he quickly supported the Bourbon restoration (when the old ruling dynasty of France returned) after Napoleon was overthrown. It was at this point under the Bourbon king Louis XVIII that Laplace was given the title of marquis, a rank of French nobility.

Laplace was criticized for his apparent political opportunism at the time, but, according to historian Roger Hahn, his abandonment of Napoleon was the result of Laplace's doubts about the wisdom of Napoleon's continued wars as well as a personal falling out between the two after the death of Laplace's daughter. In one incident recorded at the time and quoted by Hahn, Napoleon "accosted Mr. Laplace: 'Oh!, I see that you have grown thin—Sire, I have lost my daughter—Oh! That's not a reason for losing weight. You are a mathematician; put this event in an equation, and you will find that it adds up to zero." This exchange between the two men took place near the end of Napoleon's reign, as his empire was unravelling and hostile armies were advancing on France from both Spain and central Europe.

After Napoleon's ouster and nearing the age of seventy, Laplace continued to publish important scientific works. He died in 1827 at the age of eighty-one. First buried in a cemetery in Paris, his body was moved in 1888 to a family estate in the north of France near where he was born.

ALBERT EINSTEIN

In the early 1900s, Albert Einstein's theory of relativity replaced Newton's law of gravitation, which had been the foundation of modern science for more than two hundred years. Today, the theory of relativity remains a cornerstone of theoretical physics, even though there have been many developments in the field since that time. In addition to the theory of relativity, Einstein made other significant contributions to science; perhaps his most famous is the formula $E = mc^2$.

Albert Einstein delivering a lecture in Vienna in 1921

Albert Einstein was born in the German city of Ulm in 1879. Soon after, his family moved to Munich, Germany, where Einstein's father founded an electrical engineering company. This family business, run by Einstein's father and uncle, ran into financial difficulties, and after fourteen years, the family moved once again to Pavia, Italy. As a result, Einstein attended school in Munich, Germany, until the age of sixteen, when he entered a secondary school in Switzerland while his family lived in Italy. He would finish secondary school and graduate from university in Switzerland before becoming a patent clerk there. He worked as a patent clerk for seven years between graduating from university and being appointed a professor—he received his PhD from the University of Zürich during his employment at the patent office. After publishing a number of important works in 1905, Einstein quickly rose to prominence in the world of academia. He became a full professor in 1911, before becoming director of the Kaiser Wilhelm Institute of Physics in 1914. And it was in 1919 that Arthur Eddington's observation of a solar eclipse provided evidence of the theory of relativity and made Einstein a celebrity the world over.

The fact that Albert Einstein was of Jewish descent would play a large role in the course of his life. He was never an observant Jew, but his background made him a target of Nazi Germany. The rising wave of anti-Semitism in Germany before and during World War II would result in Einstein becoming an American citizen. He renounced his German citizenship in 1933 when Adolf Hitler came to power, and the same year he accepted a position teaching at the Institute for Advanced Study in Princeton, New Jersey. He would work there until his death in 1955.

Einstein used his celebrity to help other Jewish scientists who were persecuted in Germany. He lobbied other countries to accept them as refugees after Germany passed a law banning Jews from being professors. Many of these refugees would go on to make significant academic achievements, and some helped with the later American development of the atomic bomb. Einstein also supported the cause of Zionism (the political movement to establish a Jewish homeland in the biblical Holy Land). This movement eventually led to the creation of the modern state of Israel. In 1952, the relatively new state of Israel even offered him the post of president (a largely ceremonial position), though he declined the offer.

Although he was a lifelong pacifist, Einstein did play a role in the Manhattan Project, the American project to develop the atomic bomb during World War II. He signed a letter to President Franklin D. Roosevelt about the possibility of Nazi Germany developing an atomic bomb. The letter suggested that the US government "speed up" the experimental work that might result in this new type of weapon in order to counter the German threat. In spite of the fact that Einstein never worked on the Manhattan Project himself (he was deemed a security risk because of his socialist and pacifist views and not asked to participate), his letter to President Roosevelt was instrumental in the creation of the project. Einstein would come to regret his letter to President Roosevelt after Nazi Germany failed to develop their own atomic bomb, and the American atomic bombings of Hiroshima and Nagasaki killed more than 100,000 civilians.

Visualizing the Movements of the Earth, Sun, and Moon

Humans have long tried to visualize the universe. The earliest known attempts depict a flat earth with the celestial objects moving in the heavens above. But by the sixth century BCE, the Greek philosopher Pythagoras had conceived of a spherical Earth, and Aristotle later provided evidence this was the case. From this time until Copernicus, Earth was still placed at the center of the universe, and so early pictures of the solar system have the sun, moon, and planets orbiting the Earth. It was not until Galileo's discovery of the moons of Jupiter that there was widespread acceptance that celestial objects could orbit something other than Earth. After the acceptance of the Copernican model of the solar system, relatively accurate visualizations of the solar system became available. The planets were ordered correctly and their elliptical orbits around the sun were depicted, as well as the orbits of the known satellites of the individual planets. Now, with advances in technology, we have access to precise models of the solar system. In this chapter, we will look at the first written models, computer-generated models, and a physical model built on a grand scale in Sweden.

ALMANACS

Before scientists had created models of the solar system and detailed visualizations of the seasons, lunar phases, and tides, it was important for ordinary people to know when these phenomena would occur. The equinoxes and solstices were used to determine when to plant and harvest crops, and the time of the tides was essential information for sailors. While specialists were able to calculate these phenomena thousands of years ago, the ordinary person was not. As a result, what we now call almanacs were published. These books contained the dates for significant astronomical events like the equinoxes, solstices, and phases of the moon.

The first book of this type archaeologists have found dates to the time of Babylon. It contains not only predicted astronomical events but also gave advice about certain days. One day might be good to "harvest grain" or "favorable in a lawsuit." Another day might be bad for a wedding ("who marries will become old"). These translations of the tablets

Poor RICHARD improved:

BEING AN

ALMANACK

AND

EPHEMERIS

OF THE

MOTIONS of the SUN and MOON;

THE TRUE

PLACES and ASPECTS of the PLANETS;

THE

RISING and *SETTING* of the *SUN;*

AND THE

Rising, Setting *and* Southing *of the* Moon,

FOR THE

YEAR of our LORD 1 7 5 8:

Being the Second after LEAP-YEAR.

Containing also,

The Lunations, Conjunctions, Eclipses, Judgment of the Weather, Rising and Setting of the Planets, Length of Days and Nights, Fairs, Courts, Roads, *&c.* Together with useful Tables, chronological Observations, and entertaining Remarks.

Fitted to the Latitude of Forty Degrees, and a Meridian of near five Hours West from *London* ; but may, without sensible Error, serve all the NORTHERN COLONIES.

By *RICHARD SAUNDERS*, Philom.

PHILADELPHIA:

Printed and Sold by B. FRANKLIN, and D. HALL.

Benjamin Franklin published this almanac for more than twenty-five years.

are taken from an academic article published by Caroline Waerzeggers, an Assyriologist (a person who studies ancient Mesopotamia). She examined the almanacs and compared the advice to legal records from the time period. She goes on to conclude that the personal advice offered by the early almanac influenced daily life in Babylonia. She writes:

> *The most important conclusion to be drawn from our inquiry is that there is a good match between specific predictions, recommendations, and prohibitions found in the Babylonian Almanac and real-life activities recorded in the archival texts from Borsippa from the late 7th to early 5th century BC, especially in the realm of marriage.*

Almanacs remained an important tool for ordinary people, especially farmers, from the time of Babylon until the modern era. The famous founding father of the United States, Benjamin Franklin, published his own almanac between 1732 and 1759. Like other almanacs of the time (and even the present day), Franklin's almanac contained weather forecasts in addition to tide tables and the dates of the seasons and lunar phases. Although these weather forecasts may have shaped life at the time, much like the omens of Babylon, they were not based on accurate scientific predictions. However, the tables that contained the times of the tides, equinoxes, solstices, and lunar phases were both accurate and useful at the time.

Tide Tables

Tide tables are an effective means of giving information about the tides at a certain place. In the past, one had to look in an almanac to find a tide table, but now they are available online. A tide table provides the time of every high and low tide as well as the height of those tides. This height differs depending on the day due to the many variables that affect the tides. From these numbers on a tide table and a nautical chart, it is possible to determine the depth of water at any location and time. In this circumstance, a relatively low-tech solution—a list of numbers—is an efficient and easy-to-use representation of complex data.

THE SOLAR SYSTEM

Anyone who attended school in the United States is familiar with making models of the solar system. Students create basic models in class using spherical objects of various sizes and colors to represent the sun, moon, and planets. While these exercises often impress on students the massive size of the sun (often a balloon or bowling ball) versus the tiny size of the planets (usually pinheads), these sorts of models do not do justice to the vast expanse of empty space in the solar system. To truly illustrate the size of the solar system, you need to put a balloon in the middle of a field to represent the sun and walk nearly four-tenths of a mile away—the

JULY 2006

		LOW TIDE			HIGH TIDE			
		AM	Ht.	PM	Ht.	AM	Ht.	PM	Ht.

		AM	Ht.	PM	Ht.	AM	Ht.	PM	Ht.
		Sunrise 5:41		–PDT–		Sunset 8:27			
1	Sa	9:20	0.5	10:05	2.7	2:04	4.3	4:33	4.2
2	Su	9:54	1.0	11:27	2.4	3:01	3.8	5:06	4.4
3	M	10:27	1.4	----	---	4:17	3.3	5:38	4.7
4	Tu	12:39	1.8	(11:03	1.9)	5:54	2.9	6:10	4.9
5	W	1:38	1.3	(11:42	2.3)	7:36	2.9	6:45	5.2
		Sunrise 5:43		–PDT–		Sunset 8:26			
6	Th	2:26	0.6	12:29	2.6	9:02	3.0	7:21	5.5
7	F	3:09	0.1	1:19	2.8	10:05	3.2	8:01	5.8
8	Sa	3:50	–0.4	2:11	2.9	10:53	3.4	8:43	6.1
9	Su	4:30	–0.9	3:02	3.0	11:33	3.6	9:27	6.4
10	M	5:11	–1.2	3:53	2.9	(12:11	3.7)	10:12	6.6
		Sunrise 5:46		–PDT–		Sunset 8:24			
11	Tu	5:52	–1.4	4:46	2.8	(12:48	3.9)	10:59	6.6
12	W	6:33	–1.4	5:42	2.7	(1:25	4.0)	11:48	6.4
13	Th	7:14	–1.2	6:44	2.6	----	---	2:04	4.3
14	F	7:54	–0.8	7:52	2.4	12:40	5.9	2:43	4.6
15	Sa	8:35	–0.2	9:09	2.1	1:37	5.3	3:24	4.9
		Sunrise 5:49		–PDT–		Sunset 8:22			
16	Su	9:16	0.4	10:32	1.6	2:43	4.5	4:07	5.2
17	M	9:58	1.1	11:56	1.1	4:04	3.8	4:52	5.5
18	Tu	10:45	1.8	----	---	5:44	3.3	5:40	5.8
19	W	1:10	0.4	(11:38	2.4)	7:32	3.2	6:30	6.0
20	Th	2:14	–0.1	12:40	2.8	9:04	3.3	7:21	6.1
		Sunrise 5:53		–PDT–		Sunset 8:19			
21	F	3:08	–0.5	1:44	3.0	10:11	3.6	8:11	6.2
22	Sa	3:56	–0.8	2:42	3.0	10:59	3.8	8:58	6.2
23	Su	4:38	–0.9	3:34	3.0	11:39	3.9	9:43	6.2
24	M	5:17	–0.8	4:19	2.9	(12:13	4.0)	10:24	6.1
25	Tu	5:53	–0.7	5:02	2.8	(12:44	4.0)	11:03	5.9
		Sunrise 5:57		–PDT–		Sunset 8:15			
26	W	6:26	–0.5	5:44	2.7	(1:14	4.1)	11:40	5.6
27	Th	6:56	–0.2	6:28	2.7	----	---	1:42	4.1
28	F	7:25	0.2	7:16	2.6	12:17	5.2	2:11	4.2
29	Sa	7:53	0.6	8:11	2.5	12:56	4.8	2:39	4.4
30	Su	8:19	1.1	9:15	2.3	1:40	4.3	3:09	4.5
		Sunrise 6:01		–PDT–		Sunset 8:11			
31	M	8:45	1.6	10:28	2.0	2:34	3.7	3:40	4.7

3 ☽   11 ○   17 ☾   25 ●

Tide tables like this can be found online and sometimes in person at local marinas and docks.

length of six football fields—to place the dwarf planet Pluto (which would be about the size of a grain of sand). While this experiment, designed by NASA, would provide a more accurate representation of the amount of empty space in the solar system, most models simply omit the empty space so a model of the solar system can appear on a single page of a book.

However, some models do incorporate the empty space that actually makes up most of the solar system. The largest model to do so is the Sweden Solar System. It is composed of nineteen models at a scale of 1:20 million. The Globe Arena, the largest spherical building in the world, represents the sun and is located in Stockholm, the capital of Sweden. All of the inner planets (Mercury, Venus, Earth, and Mars) and the moon are also in the city of Stockholm. Because they are to scale, these inner planets range from just 9.8 inches (25 centimeters, the size of a basketball) to 25.5 inches (65 cm) in diameter, just a fraction of the size of the massive arena that represents the sun. The outer planets (Jupiter, Saturn, Uranus, and Neptune), dwarf planets (Pluto, Ixion, Eris, and Sedna), and asteroids all lie outside of Stockholm, spread across Sweden. The placement of these objects gives some insight into the vastness of the solar system—the inner planets are all in the same city as the sun while Neptune is over 140 miles (225 km) away, and the termination shock (near the end of the sun's influence over space) is almost 600 miles (966 km) away.

Many visualizations of the solar system distort more than just the amount of empty space. Often the orbits of the planets are circular (rather than elliptical), or they are elliptical to an exaggerated degree. In reality, the orbits of the planets are elliptical, but with the exception of Mercury (as well as other objects like dwarf planets and comets), they appear to be circular at a glance because they are not overtly elliptical. They do not appear oval to the naked eye—as they are presented in many diagrams that highlight Kepler's findings.

You might think that the orbits portrayed on the pages of many textbooks are also inaccurate because they are two-dimensional and, therefore, the orbits of all the planets seem to lie in the same plane. In other words, the orbits of the planets line up. But this is not an inaccuracy. In fact, the orbits of the planets do line up very precisely. If you could view the solar system from outside of it (and see the planets, which would be incredibly small in the vast emptiness of space), the planets and the sun would form an almost perfect line.

Because of the vast distances involved in modeling space and the relatively tiny size of the planets, it is impossible to create a scale model of the solar system on a single page or the screen of a computer. The planets would be too small to see. But the models that appear in books or on the web do reveal important information depending on their purpose: whether that is to show the relative size of the planets to one another or the ordering of the planets. The Laboratory for Atmospheric and Space Physics at the University of Colorado, Boulder, provides a free simulation of the orbits of the planets on their website. Although it is not to scale (the sun is not true to size), it does provide an interesting visual representation of the solar system. It is easy to see how quickly the inner planets move along their orbits compared to the outer planets (due to the effect of gravity being greater closer to the massive sun). The model also illustrates the fact that the inner planets are much closer to the sun than the outer planets—as well as much smaller than it. The elliptical orbits of the objects are also accurate, meaning the orbit of most planets appears circular but the orbits of Mercury and Pluto seem slightly elongated while the orbit of the Comet Borrelly is oval.

The Seasons

The Seasons and Ecliptic Simulator (provided free of charge online by the Nebraska Astronomy Applet Program at University of Nebraska-Lincoln) shows how the seasons change due to the rotation of the Earth around the sun. It not only portrays the Earth's orbit, but it also shows how the angle of sunlight changes for someone who is standing on the

The Einstein Cross

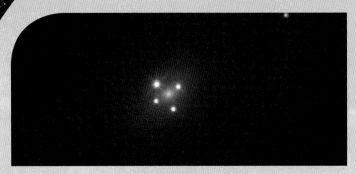

The Einstein Cross

The theory of relativity has the reputation of being difficult to understand and, perhaps, impossible for anyone but professional scientists to see clearly. However, there are a number of startling sights in the night sky that confirm Einstein's theory. One of them is named the Einstein Cross. If you look through a telescope at a distant galaxy, it appears to be surrounded by four points of light. In reality, there is only one quasar behind a massive galaxy (both the galaxy and the quasar resemble a star from Earth). Quasars are also known as quasi-stellar radio sources, and they are actually one of the brightest objects in the universe. Astronomers now think quasars are jets of energy and matter beaming out of a black hole. The light from this bright quasar is bent around the large galaxy in front of it, and this makes it appear as though there are four distinct quasars surrounding the galaxy in the middle. This is an example of gravitational lensing—when a massive object (often a galaxy) bends space-time to such a degree that the light from an object behind it is distorted. When this effect is strong enough, the object can seem to appear multiple times or simply appear to be a ring or arc of light instead of revealing its true shape.

Earth 700 miles (1,126.5 km) above the equator. The sunlight is direct twice a year (as the sun passes directly overhead), and it is relatively direct all summer. This causes the temperatures to be higher during this time (and afterwards due to seasonal lag). During the winter months, the sunlight is sharply angled, and this causes lower temperatures.

Lunar Phases

A video created by NASA and uploaded on YouTube called "Moon Phase & Libration: Moon With Additional Graphics" shows the lunar phases in high resolution and high speed. The lunar phases take place over 25 seconds rather than 29.5 days. A diagram in the top-left part of the video also shows a view of the Earth (spinning on its axis) and the moon (orbiting the Earth). From this, you can see that half of the moon is always illuminated, but it is not always the half that is visible from the Earth. This video makes it easy to see how the lunar phases result from sunlight illuminating the moon.

Eclipses

NASA's Scientific Visualization studio has a number of different videos, available online, that show the lunar eclipse of September 27, 2015. The videos allow you to see a lunar eclipse without having to wait for the right time to go outside and look at one in person. It is possible to see how, at the start of the eclipse, the moon seems to disappear. It then turns reddish before seeming to disappear again just prior to returning to its normal appearance. In a related video released

A solar eclipse viewed from the moon

by NASA, the same eclipse as it would appear to someone on the moon is portrayed. It resembles a solar eclipse from Earth, but rather than the moon blocking our view of the sun, it is Earth that does so. Astronaut Alan Bean saw such an eclipse in 1969 and said it was "a marvelous sight." The outer edges of Earth would be bathed in a red light—sunset and sunrise if one were on the ground there—and the sun would be totally obscured by Earth.

When compared to lunar eclipses, solar eclipses are extremely brief, lasting only a matter of minutes. They are also only visible from an extremely small area of Earth's surface. Another video by NASA shows the path of a solar eclipse. You can see that the shadow of the moon is only a small dot that races across the surface of the Earth. Only people who happen to be along the path of the shadow see a solar eclipse.

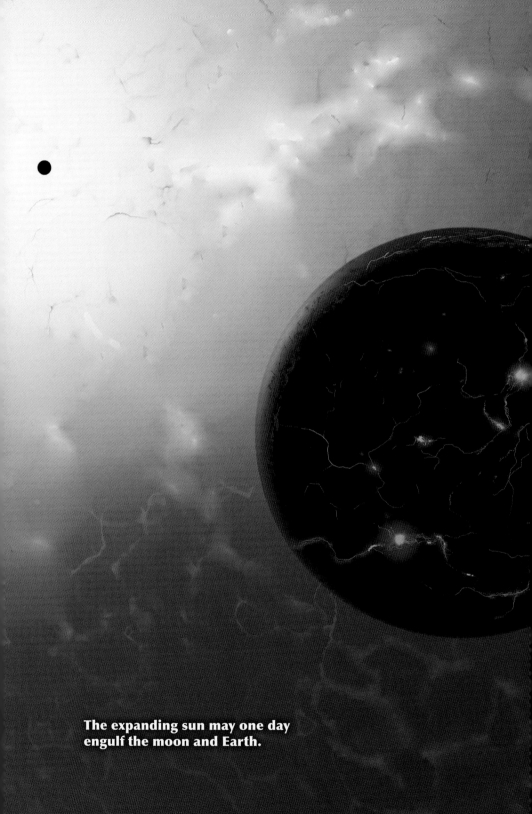

The expanding sun may one day engulf the moon and Earth.

Lunar Phases, Eclipses, and Seasons: Today and Tomorrow

The cyclical movements of Earth and the moon will gradually change in the future as the moon moves farther away from Earth and the rotation of Earth slows. This will affect life on Earth in a number of ways. Some changes will take place over just thousands of years, while others will occur millions of years in the future. In this chapter, we will examine how these changes will alter the lunar phases, eclipses, and seasons as well as all life on Earth.

CYCLIC PATTERNS AND CLIMATE CHANGE

It is now thought that cyclic patterns in the movement of Earth are one of the principal drivers of climate change over the course of thousands of years. This theory was first put forward by Serbian astronomer Milutin Milanković in the early twentieth century. But it was not until the 1970s that scientific data confirmed his suspicion and led to widespread acceptance in the scientific community. The theory is now referred to as the **Milankovitch theory** (after an alternate spelling of his name) in his honor. According to the Milankovitch theory, the varying amount of solar radiation that Earth receives is responsible for climate change. As we have seen, this also explains the changes in seasons on Earth—Earth's axial tilt leads to higher levels of solar radiation in the summer and thus warmer weather. Milanković's innovation was to apply this reasoning to variations in the movement of Earth over thousands and tens of thousands of years rather than the seasons in a single year.

Milanković identified three cycles that affect the amount of solar radiation that reaches the Earth: the axial tilt of Earth, the eccentricity of Earth's elliptical orbit, and the precession of the equinoxes.

Axial Tilt

In chapter 2, we learned that the axial tilt of Earth causes the seasons. The hemisphere that is tilted towards the sun receives more solar radiation and is warmer as a result, while the hemisphere that is tilted away from the sun is cooler. But the degree to which Earth's axis is tilted actually changes slowly, and, as you may have guessed, the greater the tilt of Earth, the more extreme the seasonal variations in temperature. The axial tilt of Earth varies between approximately 22° and 24.5° over a cycle that takes approximately forty-one thousand years to complete. Currently, it is 23.4°—close to the middle of its range. While forty-one thousand years may sound like a long time, in terms of Earth's history it is quite short. In fact, it has even changed significantly in recorded human history. Since Pytheas first measured it over two thousand years ago, it has decreased by close to 0.3°. As it continues to decrease, this will contribute to milder seasons, although its effect on Earth's climate may very well be outweighed by other factors such as climate change due to human activities.

A Changing Orbit

Earth's orbit is always elliptical, but the degree to which it is elliptical varies over the course of a cycle of approximately 100,000 years. Geographer Michael Pidwirney describes the impact this has on Earth's climate:

> *The greater the eccentricity of the orbit (i.e., the more elliptical it is), the greater the variation in solar energy received at the top of the atmosphere between the Earth's closest (**perihelion**) and farthest (**aphelion**) approach to the sun. Currently, the Earth is experiencing a period of low eccentricity. The difference in the Earth's distance from the sun between perihelion and aphelion (which is only about 3%) is responsible for*

approximately a 7% variation in the amount of solar energy received at the top of the atmosphere. When the difference in this distance is at its maximum (9%), the difference in solar energy received is about 20%.

From these figures, we can see that the elliptical orbit of Earth does have some effect on Earth's climate. While it is a misconception that the seasons result from the distance between Earth and the sun, it can be a contributing factor in long-term climate change.

The Precession of the Equinoxes

You may recall the precession of the equinoxes refers to the fact that Earth "wobbles" as it rotates on its axis. This too has an effect on Earth's climate. Right now, Earth is closest to the sun (perihelion) in January and farthest from the sun in July (aphelion). This somewhat moderates the climate in the Northern Hemisphere: during winter the sun is closer and during summer the sun is farther away. It has the opposite effect in the Southern Hemisphere. It makes the seasons more extreme because the sun is closer during the summer and farther away during the winter. But due to the precession of the equinoxes, this will gradually change over the course of a twenty-six-thousand-year cycle. In thirteen thousand years, the climate of the Southern Hemisphere

In a time-lapse photo, the North Star does not move while the other stars appear to revolve around it.

will be more moderate because the sun will be closer during the winter and farther away during the summer, and the Northern Hemisphere will have a greater seasonal variation in temperature.

Understanding Milankovitch's Theory

These three cycles all affect Earth's climate simultaneously. Their impact can also be blunted by other climactic changes that occur simultaneously with differing effects. For example, at a time when Earth's tilt is at its greatest angle (leading to more extreme seasons), Earth's elliptical orbit may be at its most circular (leading to less extreme seasons). The greatest changes occur when all three cycles align to produce a period of exceptionally cold or warm temperatures, but usually the cycles are not aligned.

It is important to remember that these cycles take place over the course of thousands of years. When you hear politicians and scientists discuss "climate change" today, they are almost always referring to the relatively recent phenomenon of climate change that results from human activities. This is often called "global warming."

GLOBAL WARMING

The average temperature of the earth has been rising gradually over the past century. While the average temperature has only increased approximately 2 degrees Fahrenheit (1.1 degrees Celsius), this change is large enough to have a significant impact on Earth's climate. Extreme weather events like floods, droughts, and heat waves have increased in frequency as a result. In the future, the sea level may rise and cause vast expanses of coastline—currently inhabited by people—to be below sea level. If the issue of global warming is not tackled, it will cause immense human suffering. Despite the fact that global warming is driven by high-income countries, people in low-income countries are expected to suffer the most as diseases spread, economic

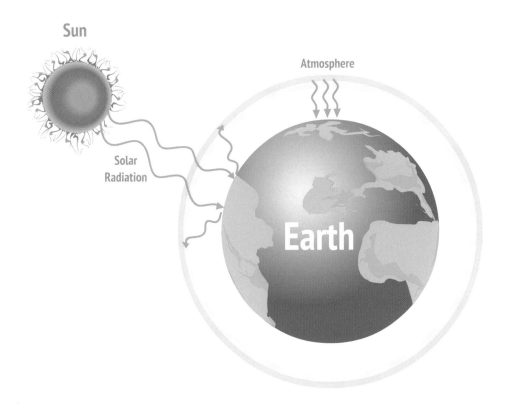

The greenhouse effect is the trapping of solar radiation in Earth's atmosphere.

growth slows, and conflicts erupt over water rights and food scarcity.

As we have seen in our discussion of seasons, the earth is warmed by solar radiation from the sun. But according to data that scientists have collected, the level of solar radiation that is reaching Earth has not changed significantly over the past century. Instead, global warming results from something that happens after solar radiation reaches the earth: the

greenhouse effect, a process in which the atmosphere of the earth traps a great deal of the sun's energy. While this process is essential for life on earth, the greenhouse effect has increased over the past century and caused the average temperature of Earth to rise as a result. This magnification of the greenhouse effect is due to rising levels of greenhouse gases—a term that refers to water vapor, carbon dioxide, and methane, among others. Greenhouse gases cause heat from the sun to remain near the surface of the earth and in the lower parts of the earth's atmosphere. Humans have caused the recent increase in greenhouse gases through a number of different activities. The most significant contributor is the burning of fossil fuels (coal, petroleum, and natural gas) for energy. This releases large amounts of carbon dioxide. But fossil fuels are not the only source of greenhouse gases. Livestock farming produces large amounts of methane (from the digestive process of the animals), and deforestation (the carbon dioxide trees store is released when they are burned or cut down).

Global warming has already caused dramatic changes to Earth's seasons. If the seasons are calculated from the lifecycle of plants rather than the equinoxes and solstices, spring is beginning earlier while autumn is beginning later as average temperatures increase and summer lengthens. According to a large scientific study released in 2006, the beginning of spring in Europe is advancing by 2.5 days each decade. This change in Earth's climate will have a sizable impact on plant and animal life. There may be some positive effects from this change: a Norwegian research group estimates the growing season there will lengthen by two months and benefit farmers. However, the majority of the effects will be negative: tropical diseases like malaria will spread; the natural range of animals will change, destroying delicate ecosystems; and polar ice will melt, causing sea levels to rise.

In the near future, global warming will continue to affect the seasons to a much greater degree than the cyclic patterns that Milankovitch's theory identifies. Global warming is a threat that humanity will have to confront as the human activities that cause it continue to occur. A positive step was recently made in December of 2015 when most of the countries of the world agreed to the Paris Agreement. This agreement seeks to limit global warming to just 1.5 degrees Celsius (2.7 degrees Fahrenheit). This is an ambitious goal (and many nations have not yet ratified the agreement), but the coordination of so many different countries is a promising development in the fight against global warming.

A CHAOTIC SOLAR SYSTEM

The question of whether the solar system is stable (and the orbits of the planets fixed) has troubled astronomers for centuries: will the planets eventually collide, pull away

Seasons on Mars

In recent years, the colonization of Mars has been a popular topic. Buzz Aldrin, an astronaut who walked on the moon in 1969, has argued before Congress in favor of a colony on Mars within twenty-five years, and Elon Musk, the billionaire CEO of SpaceX, hopes to establish a colony on Mars within the next ten years. In light of these plans, it makes sense to consider what life would look like on Mars. As you probably know already from movies, it would be necessary to wear a space suit whenever you went outside on Mars; but what would the days and seasons look like there once you were outside? A day on Mars would actually be quite similar to a day on Earth. It would last 24.6 hours—a relatively small difference from the 24 hours on Earth. But the seasons on Mars would be quite different. This is due to the highly eccentric orbit of Mars. Its orbit is more elliptical than any planet other than Mercury. As a result, the seasons of Mars are very unequal in length. Spring on Mars lasts 194 days, while autumn lasts only 142. This is because Mars's orbital speed is much higher when it is closer to the sun than it is when it is farther from the sun. As a result of this discrepancy, the seasons on Mars are more extreme than on Earth. The ice caps move to the opposite pole in the summer and winter as the sun melts whichever pole is pointed towards it. This change is so dramatic that the atmospheric pressure on Mars is 25 percent lower in the winter than it is in the summer!

from the sun, or be pulled into the sun to a fiery end? Isaac Newton's realization that the gravitational field of the planets influenced the orbits of nearby planets began the debate, but even today it is not completely settled. It is agreed that some of the orbits in the solar system are chaotic, but there is still scientific debate about other planets—like Earth. Mercury, with the most eccentric orbit in the solar system, is also the most likely suspect to cause a planetary collision in the future. While researchers estimate that there is only a 1 percent chance that Mercury will spin out of control in the next few billion years, such a scenario could spell disaster for other nearby planets like Earth and Venus. There is even the possibility that the orbits of Earth and Venus will be disrupted to the point where the two planets collide. Luckily, such an outcome is only possible billions of years from now.

However, some bodies in the solar system may change their orbits in a much shorter time frame. Researchers Gerald Sussman and Jack Wisdom used a numerical model to estimate that the orbit of Pluto may be unpredictable in just 20 million years, which is a relatively short span of time given the solar system is over 4.5 billion years old. One reason for its chaotic orbit is the fact that its orbit crosses Neptune's orbit. The two will never collide because the time period it takes the two planets to orbit the sun is regular: Pluto completes three orbits in the same time it takes Neptune to complete two. As a result, scientists know they will never

There is a chance the chaotic nature of some planetary orbits will cause a collision in the future.

collide in the course of their current orbit, but Neptune's large, nearby mass is enough to introduce significant uncertainty into Pluto's future movements.

THE EARTH AND MOON

Earth is not only subject to cyclic variations like the precession of the equinoxes, eccentricity of its orbit, and degree of its axial tilt. Earth's movements are also subject to gradual, permanent changes over millions of years. These changes will eventually cause dramatic changes in the seasons, eclipses, and lunar phases.

The spinning of Earth on its axis is gradually slowing because of the presence of the moon. As the moon causes the tides by making water on Earth's surface bulge outwards toward it, friction is created by the water dragging along the ocean floor. This friction is slowing Earth's rotation. While this change is imperceptible in our lifetime, it adds up over the course of millions of years. In fact, billions of years ago, a day on Earth lasted just six hours instead of twenty-four hours. In the future, days on Earth will continue to lengthen. Even now, leap seconds must occasionally be added because the day is slightly longer than twenty-four hours. The last leap second occurred on June 30, 2015. As Earth continues to slow, more and more leap seconds will need to be added. Some scientists have called into question whether this will be sustainable in the future, but leap seconds will remain a viable solution to this problem in our lifetime.

In addition to slowing the spin of the earth, the gravitational interplay between the tides, moon, and Earth is also causing the moon to slowly become farther and farther away from Earth. Eventually, this will dramatically effect solar eclipses as seen from Earth. In roughly 600 million years, the moon will be so far away from Earth that it cannot block Earth's view of the sun even when it is positioned between the two. This means that total solar eclipses will be a thing of the past.

This increasing distance between Earth and the moon also means that months will gradually lengthen as it takes more and more time for the moon to complete one revolution around the earth. But the length of a day on Earth will also be lengthening due to its slowing rotation. If this were to continue into the future, the length of time of a day and a month would actually become equal at roughly forty-seven days (although it is possible the sun will swallow both Earth and moon by this time). Astrophysicist Neil deGrasse Tyson explains this in the following way, "Earth's rotation will continue to slow down, and the moon will continue to spiral away until the Earth day exactly equals the lunar month. At that time, one Earth rotation will last over 1000 hours, which would require 4 million leap seconds per day. No need to panic just yet. You have over a trillion years to think about it." If this does come to pass, the moon and Earth will be tidally locked to one another. In other words, one side of Earth will always see the moon while the other will never see it. But much more importantly for life on Earth, night would be 500 hours of darkness, and plants and animals would have to live radically different lives as a result.

AN EXPANDING SUN

One day, life will cease on Earth as the solar system gradually changes. We do not yet know how this will happen. Will an asteroid collide with our planet? Will the sun expand and heat Earth until it is a fiery ball of molten rock? Or will humanity destroy itself before any astronomical event even occurs?

While we cannot predict whether or not a celestial object will collide with Earth in the distant future, we do know for a fact that the sun will expand and destroy life as we know it. In about five million years, the sun will run out of the hydrogen in its core, and hydrogen currently powers the nuclear fusion taking place there. As nuclear fusion takes place outside the inner core, the sun will become a red giant. Swelling to more than two hundred times its current size, it will fill the space that Mercury, Venus, and Earth currently lie in.

Whether Earth will be pushed outwards by gravity or simply perish in the flames is still an open question. What is clear is that the orbits of Earth and the moon (if they still exist) will be radically different at this time because of the expansion of the sun. The seasons, eclipses, and lunar phases will also differ due to this change in orbit. Earth will likely be uninhabitable, but five million years from now it is possible that humans will be able to leave Earth en masse.

The movements of the moon, sun, and Earth will largely stay the same for millions of years.

CYCLES OF OUR LIFE

In our lifetime, the orbits of the moon and Earth will continue without any perceptible changes. The sun too will remain in its current state and will not begin to expand for millions of years. The cycles of the lunar phases, eclipses, and seasons will carry on—driven by the effects of gravity on the motions of the celestial objects. Eclipses will continue according to Saros cycles (first noticed by the Babylonians) as Earth and the moon continue to orbit. The lunar phases will light up the night sky due to our view of the orbiting moon illuminated by the sun. The seasons will come and go as a result of the tilt of Earth's axis. Yet, over the course of thousands of years, minor changes in the movements of Earth and the moon will drive climate change. But this slow change in Earth's climate may be overtaken by the dramatic effects of global warming unless it is stopped in the near future.

As we have seen, it took thousands of years for our current understanding of these cyclic patterns to evolve. A diverse line of influential thinkers shaped our understanding of the universe. Often opposed by political forces or doubted by their peers, these thinkers fought to expand the bounds of human understanding. When we look at the world around us, we can better understand it because of their contributions to science. The effects of the complex movements of Earth and the moon are visible to us in the lunar phases, eclipses, seasons, and tides. These phenomena have shaped human history in the past and will continue to do so as long as there is life on Earth.

Glossary

aphelion The point in an orbit when the object is farthest from the sun.

axial tilt The degree to which Earth's axis is tilted relative to its orbit around the sun.

equinox The two days of the year when Earth's equator crosses the central point of the sun, meaning day and night are roughly the same length.

Copernican Revolution The radical change in scientific understanding from a geocentric model to a heliocentric model; it began with the work of Copernicus and ended with Newton's undeniable proof of the heliocentric model.

epicycles Theorized small circular movements that celestial bodies make in the course of their large circular orbits; though present in both the Ptolemaic system and Copernican model of the solar system, they do not exist.

geocentric model The model of the solar system that places Earth at the center of the universe with the sun, moon, and planets orbiting it.

greenhouse effect The primary driver of global warming whereby greenhouse gases trap energy in Earth's atmosphere.

Gregorian calendar The calendar that is used today in the United States and across most of the world.

heliocentric model The model of the solar system that places the sun at the center of the solar system with the planets orbiting it.

law of universal gravitation A mathematical description of the effects of gravity developed by Isaac Newton; it has been replaced by Einstein's theory of relativity, but it is still an accurate approximation of gravity.

lunar eclipse When Earth blocks sunlight from reaching the moon, causing it to dim or take on a reddish hue.

Milankovitch Theory The theory that long-term climate change is due to cyclical variations in the movement of Earth.

perihelion The point in an orbit when the object is closest to the sun.

precession of the equinoxes The phenomenon whereby the position of the stars at the time of equinox changes slowly over time; one cycle takes approximately twenty-six thousand years to complete.

seasonal lag The delay between the amount of solar radiation that Earth receives and changes in seasonal temperatures; it is due to the fact that it takes time for Earth to heat or cool.

solar eclipse When the moon obstructs Earth's view of the sun.

solar radiation Energy that is emitted from the sun; it includes both visible light and invisible energy such as UV rays.

solstice The longest day (summer solstice) and shortest day (winter solstice) of the year.

stability of the solar system The theory that the orbits of the planets are stable and therefore will not be radically different in the future; historically, this was an important question in astronomy.

tidally locked A celestial body is tidally locked when it completes one rotation on its axis in the same time it takes to complete an orbit; this results in one side always facing the object it orbits.

Further Information

Books

Cunliffe, Barry. *The Extraordinary Voyage of Pytheas the Greek*. London: Penguin Books, 2003.

Isaacson, Walter. *Einstein: His Life and Universe*. New York: Simon & Schuster, 2008.

Sobel, Dava. *A More Perfect Heaven: How Copernicus Revolutionized the Cosmos*. London: Walker Books, 2012.

Websites

Isaac Newton
http://www.history.com/topics/isaac-newton
This website from the History Channel features an article about Isaac Newton's life and scientific discoveries that includes a number of videos about related topics like Copernicus and the scientific revolution.

Museo Galileo
http://www.museogalileo.it/en/explore/virtualmuseum.html
The Museo Galileo, located in Florence, Italy, allows you to visit their museum virtually. There are hundreds of videos

relating to the life and work of Galileo, and every object owned by the museum is described and pictured.

NASA's Scientific Visualizations Studio
https://svs.gsfc.nasa.gov/index.html
This website showcases the branch of NASA that provides educational visualizations concerning various topics like the movements of the planets, lunar phases, and climate change. It is updated twice a week with new material.

Videos

Introduction to Gravity
https://www.khanacademy.org/science/physics/centripetal-force-and-gravitation/gravity-newtonian/v/introduction-to-gravity
This Khan Academy video discusses the force of gravity and the mathematics of Newton's law of gravitation.

Space Time by PBS
https://www.youtube.com/channel/UC7_gcs09iThXybpVgjHZ_7g
PBS's Space Time YouTube channel focuses on space and related topics like gravity and the theory of relativity.

Bibliography

Amundsen, Bård, and Thomas Keilman. "Climate Change to Lengthen Growing Season." The Research Council of Norway. September 10, 2012. http://www.forskningsradet.no/en/Newsarticle/Climate_change_to_lengthen_growing_season/1253980646419.

Bryner, Jeanna. "Long Shot: Planet Could Hit Earth in the Distant Future." *Space.com*. June 10, 2009. http://www.space.com/6824-long-shot-planet-hit-earth-distant-future.html.

Deparis, Vincent, Hilaire Legros, and Jean Souchey. "Investigations of Tides from the Antiquity to Laplace." In *Tides in Astronomy and Astrophysics*, edited by Jean Souchey, Stéphane Mathis, and Tadashi Tokieda, 31–82. Springer, 2012.

Einstein, Albert. "On the Influence of Gravitation on the Propagation of Light."*Annalen der Physik* 35: (1911): 898–908.

Foust, Jeff. "Musk Plans Human Mars Mission as Soon as 2024." *SpaceNews*. June 2, 2016. http://spacenews.com/musk-plans-human-mars-missions-as-soon-as-2024/.

Holton, Gerald James, and Stephen G. Brush. *Physics, the Human Adventure: From Copernicus to Einstein and Beyond*. New Brunswick: Rutgers University Press, 2001.

Hoskin, Michael. *The Cambridge Illustrated History of Astronomy*. Cambridge: Cambridge University Press, 1996.

Machamer, Peter. *The Cambridge Companion to Galileo*. Cambridge: Cambridge University Press, 1998.

Machamer, Peter. "Galileo Galilei." In Stanford Encyclopedia of Philosophy. Stanford University, 2014. Retrieved June 13, 2013. http://plato.stanford.edu/entries/galileo.

Menzel, Annette, Tim H. Sparks, Nicole Estrella, Elisabeth Koch, Anto Aasa, Rein Ahas, Kerstin Alm-Kübler et al. "European Phenological Response to Climate Change Matches the Warming Pattern." *Global Change Biology* 12, no. 10 (2006): 1969–1976.

NASA. "Interplanetary Seasons." Retrieved June 2, 2016. http://science.nasa.gov/science-news/science-at-nasa/2000/interplanetaryseasons/.

Pidwirney, Michael. "Causes of Climate Change." *Fundamentals of Physical Geography, 2nd Edition*. April 13, 2010. http://www.physicalgeography.net/fundamentals/7y.html.

Sussman, Gerald J., and Jack Wisdom. "Numerical Evidence that the Motion of Pluto is Chaotic." *Science* 241, no. 4864 (1988): 433–437.

Tyson, Neil deGrasse. "The Tidal Force." Hayden Planetarium. November 1, 1995. http://www.haydenplanetarium.org/tyson/read/1995/11/01/the-tidal-force.

Waerzeggers, Caroline. "Happy Days: The Babylonian Almanac in Daily Life." In *The Ancient Near East, A Life!*, edited by Tom Boiy, Joachim Bretschneider, Anne Goddeeris, Hendrik Hameeuw, Greta Jans and Jan Tavernier, 653–644. Leuven: Peeters Publishers, 2012.

White, Michael. *Isaac Newton: The Last Sorcerer*. New York: Basic Books, 1999.

Index

Page numbers in **boldface** are illustrations. Entries in **boldface** are glossary terms.

agriculture, 5
aphelion, 87–88
Aristotle, 14, 16, 71
atomic bomb, 68–69
axial tilt, 32, 53, 86–87, 96

Babylonian, 8, 13–14, 47, **48**, 49–51, 60, 74, 100
black holes, 45

calendar, Julian, 37, 39, 57
Catholic Church, 56–57
Copernican Revolution, 21, 24
Copernicus, Nicolaus, 19–21, 23, 27, 53, 56, 59, 71
crops, 5, 13, 72

Eddington, Arthur, 38, 67

Einstein, Albert, 7–8, 31, 38, 45, 47, 58–59, 65–69, **66**, 80
epicycles, 23
equinox, 17–19, 33–34, 37, 72, 74, 86, 88, 92, 96

Galileo, 7, 24–25, 27, 53–57, **54**, 59, 71
geocentric model, 16, 20, 23–24, 27, 56, 63
global warming, 90–93, 101
gravity, 6, 8, 24–25, 27, 29–32, 35, 38, 45, 57, 63, 79, 99–100
greenhouse effect, 91–92, **91**
Gregorian calendar, 33–35, 37, 39, 41

heliocentric model, 16, 20, 23–25, 27–28, 53, 56–57, 59, 63
hemisphere, 32–34, **32**, 86, 88–89

Kepler, Johannes, 21, **21**,
    23–25, 27–28, 31, 57, 63, 78

Laplace, Pierre–Simon, 61–65,
    62
law of universal gravitation, 8,
    29–30, 44
leap year, 7, 34–35, 37, 39
lunar eclipse, 7, **19**, 22, 42, 44,
    81–83
lunar phases, 5–6, 8–9, **10**,
    19, 20, **26**, 39–42, 44,
    49, 72–74, 81, 85, 96, 99,
    100–101

Milankovitch Theory, 86, 89, 93
moon, Earth's, 5–7, 9, 11–16,
    22, 24–25, 27, 29–31, 38–
    44, **43**, 47, 49, 53, 55–56,
    63, 71–72, 75, 77, 81–83,
    85, 94, 97–101, **100**

Napoleon, 61, 64–65
NASA, 8, 77, 81–83
navigation, 13, 18
Newton, Isaac, 7–8, 24–25,
    27, **28**, 29–31, 38, 44–45,
    57–61, 63, 65, 93

orbit, 86–89, 93–96, **96**,
    99–100

perihelion, 87–88
precession of the equinoxes,
    18019, 86–88, 96
Ptolemy, 16, 20
Pytheas, 16, 18, **46**, 51–53,
    63, 87

seasonal lag, 34, 81
seasons, astronomical, 33, 35
seasons, meteorological, 33
sidereal month, 41
solar eclipse, **4**, 6, 38, 42–44,
    43, 67, 82–83, **82**, 97
solar radiation, 33, 86, 91
solstice, 33–34, 72, 74, 92,
stability of the solar system, 61
Stonehenge, 12
synodic month, 41

telescope, 24, 55, 80
theory of relativity, 8, 31, 45,
    58, 65, 67, 80
tidally locked, 41, 98
tides, 6–7, 16, 19–20, 25, 31,
    53, 63–64, 72, 74–75, **76**,
    97, 101

About the Author

Derek L. Miller is a writer and educator from Salisbury, Maryland. He has taught students of all ages in both the United States and South Korea. He is the author of *The Economy in Contemporary Africa* and *Health in Contemporary Africa*.

In his free time, Miller enjoys hiking, playing chess, and traveling with his wife. He would like to see the Sweden Solar System for himself one day.

523 MIL   Y

Miller, Derek L.,

Earth, Sun, and Moon

JUL 1 9 2017

MARY JACOBS LIBRARY
64 WASHINGTON STREET
ROCKY HILL, NJ 08553